Feeding and Care of the Horse

Feeding and Care of the Horse

LON D. LEWIS DVM, PhD
Clinical Nutritionist
Department of Clinical Sciences
Colorado State University
Fort Collins, Colorado

LEA & FEBIGER   Philadelphia

Lea & Febiger
600 Washington Square
Philadelphia, PA   19106
U.S.A.

Library of Congress Cataloging in Publication Data

Lewis, Lon D.
  Feeding and care of the horse.

  Bibliography: p.
  Includes index.
  1. Horses—Feeding and feeds. I. Title
SF285.5.L48   636.1'084   81-8137
ISBN 0-8121-0803-5   AACR2

Published in Great Britain by Bailliere Tindall, London

PRINTED IN THE UNITED STATES OF AMERICA

Print No. 3

Preface

Nothing is of greater importance in maintaining the health and productivity of an animal than proper feeding and the utilization of good quality feedstuffs. In order to feed an animal in the best possible manner, you must know something about the feeds available. You should be familiar with the different types of feeds, how to assess their quality, their advantages and disadvantages, what nutrients they contain, their function, and the utilization of these nutrients by the animal. This book is designed to further that end. It begins with a discussion of nutrients and feedstuffs, then goes to a description of how to formulate rations which will meet, or how to determine if a ration meets, the animal's nutritional requirements for optimal health and productivity. The horseman and veterinarian should have a good knowledge of these areas to ensure good feeding programs for the horse, as discussed in the last part of the book.

This book is intended for the benefit of all who own, care for, or care about horses, ranging from the 4-H club member and pleasure horse owner to the professional horseman and veterinarian. To serve this wide range of interest, knowledge, and expertise about horses, an extensive glossary has been included. Throughout the text, words defined in the glossary are written in italics the first time they appear in each paragraph. While some chapters, such as the one on epiphysitis and contracted flexor tendons in the growing horse, will be of most use to the veterinarian, the book contains information of which all horsemen should be aware.

This book is not intended as a work that is read once and put aside, but rather as a reference book to which the horseman and veterinarian can frequently refer in order to answer the multitude of questions and concerns that arise daily in feeding and caring for the horse.

It is the hope of the author that the information presented here will be of value in ensuring and enhancing optimal health and productivity of that noble animal that we are all so fond of, the horse. It is for this reason that this book was written. To fulfill this purpose, the information must be accurate and presented in a readable, and usable manner. To ensure accuracy, an extensive search of the literature on the topic was made, and the book reviewed by seventeen of the most knowledgeable people in the horse industry. These include seven veterinarians, three nutritionists, one equine reproductive physiologist, and six people involved in other aspects of the horse industry. The book will serve its purpose, however, only if you, the reader, put the information to use in the feeding and care of your horses.

Fort Collins, Colorado   Lon D. Lewis

Acknowledgements

This book and its contents were thoroughly reviewed for accuracy and clarity by numerous individuals whom the author gratefully acknowledges and thanks. They include the following: Drs. James L. Voss, Robert K. Shideler, and Simon A. Turner, all outstanding equine veterinarians at Colorado State University; Gordon R. Wooden, B.S., M.S. (Consulting Animal Nutritionist, Solvang, California) and Harold F. Hintz, Ph.D. (Professor of Animal Science, Cornell University), two of the best equine nutritionists; the entire Equine Research Committee of the American Quarter Horse Association, including previous members who were active during the planning stages of this book, namely; G. Marvin Beeman, D.V.M. (Littleton Large Animal Clinic, Littleton, Colorado); Charles W. Graham, D.V.M. (Elgin, Texas); Virginia E. Hyland (Lake Hughes, California); Don Jones (Kerrville, Texas); Stephen "Tio" Kleberg (King Ranch, Kingsville, Texas); C. A. "Chuck" Lakin (Lakin Feed and Cattle Company, Tolleson, Arizona); Kent B Linebaugh, LL.B. (Salt Lake City, Utah); Dr. Monte R. Musgrove (Olney, Illinois); Don Pabst (Atherton, California); Robert Q. Sutherland (Kansas City, Missouri); Robert Tashjian, V.M.D. (New England Institute of Comparative Medicine, West Boylston, Massachusetts); Dr. W. M. "Bill" Warren (Executive Director, Santa Gertrudis Breeders International, Kingsville, Texas); and Barry Wood, D.V.M. (Carmel, Indiana); American Quarter Horse Association Staff Members Harold G. Harms, Treasurer (Amarillo, Texas) and Mrs. Betty Nix, Administrative Assistant (Amarillo, Texas) also assisted the Equine Research Committee.

Special thanks is given to Dr. B. W. Pickett (Head, Animal Reproduction Laboratory, Colorado State University) for his encouragement to write this book and his help in doing so.

Illustrations are by Mr. Thomas O. McCracken, and many of the photographs are by Mr. James G. Bolick, both of Colorado State University. Some of the photographs were obtained from faculty of the

Colorado State University Veterinary Teaching Hospital. Many of these pictures they had taken; for others the source is not known, making acknowledgement for their use impossible. Last, but certainly not least, gratitude is expressed to Mrs. Sandra K. Swets for converting the author's scribblings into readable information, and to Miss Lynne Kesel for her excellent editing.

L.D.L.

Special Acknowledgement

The author wishes to acknowledge the encouragement and assistance of the American Quarter Horse Association in the publication of this book. Not only was the manuscript helpfully reviewed by members of the Association's Equine Research Committee, but the Association itself officially endorsed the effort in the interest of improving the quality of feeding and caring for horses, regardless of breed.

American Quarter Horse Association

2736 West Tenth Avenue
Amarillo, Texas 79168
(806) 376-4811

Formed in 1940, the American Quarter Horse Association is the world's largest equine registry, with more than 1.7 million horses registered. The purpose of AQHA is to collect, record, and preserve the pedigrees of Quarter Horses; to publish a stud book and registry; and to stimulate any and all other matters such as may pertain to the history, breeding, exhibiting, publicity, sale, or improvement of this breed. In support of these objectives, the American Quarter Horse Association founded a research program in 1960. Since then, over $1 million has been given to conduct research on selected equine problems.

Contents

THE AUTHOR

Dr. Lon D. Lewis grew up on a ranch near Harrison, Nebraska, and attended school there and in nearby Lusk, Wyoming. He grew up using horses for ranch work and pleasure. He and Nancy Lee Brunk were married in 1964 and have two boys, Bart Allen (1968) and Corey Lee (1972). He received a B.S. in chemical engineering from the University of Wyoming, and a Doctorate in Veterinary Medicine from Colorado State University in 1967. After several years of veterinary practice in Martin, South Dakota, he returned to Colorado State University where he received a Ph.D. in physiology. After two years of research at the University of Lund Medical School in Lund, Sweden, and in the Department of Internal Medicine at the Southwestern Medical School, University of Texas, he returned to Colorado State University in 1974 to fill the Mark L. Morris chair in clinical nutrition.

As a clinical nutritionist in the Department of Clinical Sciences, College of Veterinary Medicine and Biomedical Sciences, he conducts research, teaches, and provides extension and consultation services to animal owners, veterinarians, and students in the areas of feeding and management, and in nutritional, metabolic, and gastrointestinal diseases. He is the author of many scientific journal articles and numerous chapters in several veterinary textbooks. He has been invited to present papers on equine nutrition at annual meetings of the American Veterinary Medical Association, Canadian Veterinary Medical Association, American Association of Equine Practitioners, Western Canadian Association of Equine Practitioners, American Quarter Horse Association, Argentine Congress on Veterinary Sciences, numerous state veterinary medical associations, and various horse groups' meetings. He is past president of the American Comparative Gastroenterology Society and the Colorado State University Nutrition Institute. In 1980, he was awarded the American Feed Manufacturer's Association award by the American Veterinary Medical Association Council on Research for original research work on nutrition, physiology, and disease, published in recognized scientific journals during the preceding five years.

Chapter 1
Nutrients

*Nutrients** for the horse in order of their importance and the amount needed are (1) water, (2) those used for *energy*, (3) *protein*, (4) calcium, (5) phosphorus, (6) *vitamin A*, and (7) salt. The *ration* should always be evaluated in this order. First, ensure that there is adequate water and feed to meet the animal's energy needs. Then ensure that the ration contains adequate quantities of the other nutrients to meet the horse's requirements (Appendix Table 2). The horse may suffer from a variety of diseases or problems as a result of deficiencies or excesses of nutrients other than these seven. However, under normal conditions, using average or better quality feeds, it is generally not necessary to consider other than these seven nutrients in a feeding program. In some areas selenium must also be considered.

WATER

The amount of water needed by the horse in a cool environment at rest is 0.5 to 0.6 gal per 100 lbs of body weight per day (42 to 50 ml/kg). The amount needed and consumed will vary, depending upon the moisture content of the feeds eaten, environmental temperature and humidity, exercise, and lactation. For example, at an environmental temperature of 0°F (−18°C), horses will consume 1 qt of water per lb of dry feed eaten (2 L/kg), whereas at 100°F (38°C), they will consume 1 gal of water per lb of dry feed intake (8 L/kg). For practical purposes, **always ensure that adequate, good quality water is easily available free-choice.**

The only exception to this *free-choice* rule is that after exercise, the horse should be cooled off before being allowed to drink as much it wants. Cold water given to a hot, tired horse may cause *colic* and *founder*. However, during exercise the horse should be allowed to drink as frequently as practical, and as much as it wants. After prolonged exercise, allow the horse to graze or eat hay, and rest for 30 to 90 minutes before watering. After watering, *grain* may be fed.

---

*Words in italics are defined in the glossary.

Fig. 1–1 (A,B). Automatic waterers. They should be cleaned frequently and placed away from the feedbunk to prevent them from becoming contaminated with feed and hay as shown in A. A double waterer, as shown in B, may be placed between two stalls.

The first effect of inadequate water intake is decreased feed intake, followed by decreased physical activity and ability to perform work. Horses may have inadequate water intake if water is poorly accessible (Fig. 1–1), frozen over, too warm, too cold, or of poor quality, or if electric heaters with wiring problems cause the animal to be shocked when attempting to drink.

The recommended upper safe level of water contaminates is given in Tables 1–1 and 1–2. The amount of total dissolved solids (TDS), as given in Table 1–2, provides a useful overall index to the suitability of a water supply for livestock use. **TDS is the single most reliable parameter by which water quality can be evaluated.** However, contaminates in excess of those given in Table 1–1 may also make the water unsuitable for animal consumption, at least over any long period of time. Some species of blue-green algae, which grow on water, may

TABLE 1–1
UPPER SAFE LEVEL OF WATER CONTAMINATES RECOMMENDED[79]

Contaminate	Upper Safe Level Recommended*
Arsenic	0.2
Cadmium	0.05
Calcium	500
Chloride	3000
Chromium	1
Copper	1†
Fluoride	6
Hardness	200
Iron	0.3
Lead	0.1
Magnesium	125
Mercury	0.01
Nitrate	200‡
pH	6.0 to 8.5
Potassium	1400
Selenium	10§
Sodium	2500
Sulfate	250
Total Dissolved Solids (TDS) (See Table 1–2)	6500
Zinc	15‖

*All values are given in parts of contaminate per million parts of water (ppm), except selenium, which is in parts per billion (ppb), and pH. For conversion to other units, see Appendix Table 8.

†More copper than this isn't toxic but causes a bad taste.

‡High nitrate levels in water occur most commonly as a result of fecal contamination (see Chapter 4). Excessive bacteria may be present in high-nitrate-containing water.

§Although chronic selenium toxicity has been reported as a result of consumption of water containing 0.5 to 2 ppb, levels below 10 are not generally considered toxic.

‖High zinc levels may occur where galvanized pipes are connected to copper. This results in electrolysis, releasing zinc from the galvanized pipes into the water.

TABLE 1–2
A GUIDE TO THE SUITABILITY OF WATER FOR LIVESTOCK[79]

TDS (ppm)*	Suitability and Effect
1000–3000	Satisfactory for all livestock and poultry. May cause mild and temporary diarrhea in livestock not accustomed to it, but should not affect their health or performance.
3000–5000	Should be satisfactory for livestock, although it might cause temporary diarrhea, or be refused at first by animals not accustomed to it.
5000–7000	Can be used with reasonable safety for livestock. May be well to avoid water approaching the higher level for pregnant or lactating animals.
7000–10,000	Unfit for poultry and swine. Considerable risk may exist in using this water for pregnant, lactating, or young animals, or for any animals subjected to heavy heat stress or water loss. In general, the use of this water should be avoided, although animals other than those described above may subsist on it for long periods.
Over 10,000	Not recommended for use for any animal under any conditions.

*Total dissolved solids, total soluble salts or salinity in the water in ppm or mg/L.

result in poisoning; therefore, water with heavy algae growth should be avoided.

The total dissolved solids value is the sum of the concentration of all constituents dissolved in water. The term "salinity" as applied to fresh water is often used synonymously with TDS. Another term used to describe water quality is total alkalinity. It is not as good an indication of water quality as is TDS. Total alkalinity is the sum of the alkali metals, which are primarily sodium and potassium, but may also include lithium, rubidium, cesium, and francium. The hydroxides of these metals are akaline, i.e. in water they neutralize acids. The total alkalinity of water is always less than that water's TDS or salinity, since TDS and salinity include the sum of the concentration of all constituents dissolved in water, and total alkalinity includes only the sum of the alkali metals. Salinity and TDS should not be confused with hardness. Highly saline waters may contain low levels of the *cations* responsible for hardness. Water "hardness" indicates the tendency of water to precipitate soap or to form a scale on heated surfaces. Hardness is generally expressed as the sum of calcium and magnesium reported in equivalent amounts of calcium carbonate. Other cations, such as strontium, iron, aluminum, zinc, and manganese, also contribute to hardness. Hard water containing high levels of calcium and magnesium has been implicated as a cause of urinary calculi, because these minerals are important components of the calculi. However, numerous studies refute this.

ENERGY

Of the feed ingested, 80 to 90% is needed to provide the animal's *energy* needs. After water, the *nutrient's* needed to provide energy are by far the most important. There are three sources of dietary energy: *carbohydrates*, *fats*, and *proteins*. The major carbohydrates in feeds are sugars, starch, and *cellulose*. Carbohydrates in feeds are commonly referred to as crude *fiber* and *nitrogen-free extract (NFE)*. Crude fiber is primarily cellulose. It is the most poorly utilized source of energy. Much of the crude fiber ingested is excreted in the feces. Thus, the higher the fiber content of a feed, the lower the amount of energy that feed will provide, and the poorer the quality of the feed. Nitrogen-free extract is readily utilized and provides most of the dietary energy in the horse's *ration*.

Fats may be referred to as *ether extract*, lipids, or *oils*, and are also readily utilized. Although most *rations* contain only 2 to 6% fat, the horse can utilize as high as 16% in the total ration and 30% fat in the *concentrate* without adverse effects.[27] Higher levels decrease feed *palatability* and cause loose stools. Fats, particularly those containing high amounts of unsaturated fatty acids, such as any of the plant or vegetable oils, may give the horse a glossier hair coat. Many commercially available coat conditioners and *supplements* contain fats for this purpose. The same effect can be achieved by adding 1 to 2 oz (30 to 60 ml) of any of the plant or cooking oils to the horse's ration twice a day. Fats and oils have the added benefit that they may help the horse shed earlier in the spring. Frequent grooming is always beneficial. Fats may also be added to a concentrate mix to (1) increase its *energy* density, (2) decrease dust, (3) lessen wear and tear on feed mixing equipment, and (4) as a binder for pelleting or to prevent fine material such as *mineral* supplements from sifting out. The fat present in feeds may become rancid during storage, particularly at high temperatures and humidity. Rancid fats, or feeds containing them, should not be fed. Rancid fats are less palatable, do not provide the fatty acids needed, and prevent utilization of some of the *vitamins*. If they are consumed over a period of time, the major observable effect is a dry, lusterless hair coat. Fats provide 2.25 times as much energy as an equal weight of *carbohydrates* or *proteins*. If there is inadequate intake of other sources of energy, or more protein is ingested than needed to meet the horse's protein requirements, protein is used for energy. The effects of excessive protein intake are described in the following section on protein.

The *energy* that a feed will provide is expressed as *calories, therms* or percent *TDN*. In physics and chemistry, 1 calorie is the amount of energy necessary to raise the temperature of 1 g of water 1°C. However, this term is never used in nutrition. In nutrition, 1 calorie, whether written as *Calorie*, calorie, kcalories, or *kilocalories*, is equal to 1000 of the calories used in physics and chemistry. Although kilocalorie or kcalories are the correct terms to use in nutrition, calorie and Calorie are frequently used. One therm equals one *megacalorie (Mcal)* and both equal 1000 kilocalories *(kcal)*. *Total digestible nutrients* or TDN is also an energy term frequently used in nutrition. One pound of TDN is approximately equal to 2000 kcal of digestible energy (1 kg of TDN = 4400 kcal). When calculating energy intake, or the amount of feed necessary to meet the animal's energy requirements, any of the energy terms may be used. Of course, the same units must be used for the feed content and for the animal's needs.

In general, animals will eat the amount of feed needed to meet their *energy* needs if feed is available and their stomachs will hold that amount. The maximum amount of *air dry* feed that the mature horse can eat is equal to 3% of its body weight per day. For the 1000-lb horse, this means it is capable of eating 30 lbs of air dry feed daily (or for the 500-kg horse, 15 kgs of feed daily). If the horse is accustomed to the *ration*, it will not consume this much if its energy needs are met with less. Highly palatable feeds are the most common cause of overeating, which over a period of time will result in obesity. If a horse gains access to highly palatable feeds such as *grain*, it is likely to consume large quantities at one time, which can result in diarrhea, *colic*, or *founder* (see Chapter 4).

Guidelines on the amount of feed necessary to meet the horse's *energy* requirements are given in the feeding programs described and in Appendix Tables 2 and 3. These amounts are given in lbs of feed needed per 100 lbs of body weight per day (kg/100 kg/day). Weighing the feed in the container normally used to measure the feed will aid in calculating the correct daily needs. The approximate weight of a quart of most feeds, or the amount contained in a 1-lb coffee can, is 1 lb of oats or linseed meal, 0.5 lbs of bran, beet pulp or alfalfa meal, and 1.5 lbs. of other *cereal grains* or protein supplements (1 L, or a 1-kg coffee can, holds 1 kg of oats, 0.5 kg of bran, and 1.5 kg of other cereal grains).

The weight of the horse may be estimated using a horse-weight tape (Fig. 1–2). This is used to measure the girth and is marked in pounds (or kg) of body weight equal to that girth measurement for the average horse. The horse's weight can also be estimated from a girth measurement using the information in Appendix Table 4.

PROTEIN

Protein consists of many *amino acids* "hooked" together. Different types of protein consist of different combinations and numbers of amino acids. An analogy would be that amino acids may be considered similar to the letters in the alphabet and proteins similar to words.

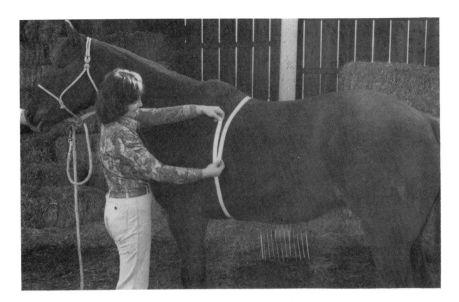

Fig. 1–2. Estimating the horse's weight from measurement of the girth. A weight tape marked in pounds of body weight (which for the average horse correlate well) is available from some feed stores and tack shops. A tape measure and the information given in Appendix Table 4 may be used instead.

Different words consist of different combinations of letters, just as different proteins consist of different combinations of amino acids. Amino acids resemble *carbohydrates* and *fats* in that they all contain many carbon molecules "hooked" together, with hydrogen and oxygen attached to the carbon. Carbon is oxidized in the body to produce *energy*. However, amino acids also contain nitrogen. Most proteins contain about 16% nitrogen; therefore, the protein content of a feed is estimated by determining its nitrogen content and dividing this amount by 0.16. The value obtained is the crude protein content of the feed. The proteins in most feeds for the horse are about 70 to 75% digestible. Thus, a feed containing 1.6% nitrogen would contain 10% crude protein (1.6% ÷ 0.16), and about 7.0% digestible protein (10% crude protein × 0.70 digestible). Both crude protein and the digestible protein content of a feed may be determined by laboratory analysis.

Generally, either crude protein or digestible protein may be used, as long as the same one is used in comparing the protein content of the *ration* to the animal's protein requirements. Crude protein is used most commonly and is therefore used throughout this book. However, digestible protein values should be used if it is suspected that the feed may have undergone excessive heating. Excessive heating of a feed will decrease its *protein* digestibility. This may occur as a result of improper processing or inadequate drying prior to storage.

There are 22 different *amino acids*. Although all of them are needed for synthesis of body *protein,* some can be produced in body tissues and do not need to be supplied in the feed or absorbed from the intestine. These are referred to as nonessential amino acids, while those that must be provided in the *ration,* or synthesized by bacteria in the intestinal tract, are called essential amino acids. Essential and nonessential, therefore, indicate whether the amino acid must be absorbed from the intestinal tract. Proteins composed of a high proportion of essential amino acids are referred to as high-quality proteins. Those containing a high proportion of nonessential amino acids are low or poor-quality proteins.

For the mature horse, the quality of *protein* is unimportant, because bacteria in the horse's intestinal tract produce all the *amino acids* needed. After ingestion of a poor-quality protein, these bacteria convert it to essential amino acids. However, a greater amount of *lysine,* an essential amino acid, is needed for growth than the bacteria are capable of producing and than is present in many feedstuffs. Two other essential amino acids, methionine and tryptophan, are present in low quantities in *cereal grains.* The most limiting amino acid in the growing horse's *ration,* however, is generally lysine. At least 0.65% lysine is needed in the total *air dry* ration for optimal growth and development of the horse. The effects of inadequate lysine are well illustrated in the study shown in Table 1–3 in which (1) soybean meal, (2) cottonseed meal, and (3) cottonseed meal fortified with lysine were used as the protein *supplements* in the ration. Although all three rations had an identical protein content, when cottonseed meal without added lysine was fed, growth rate was slower and more feed was required.

The percentage of *lysine* in the *protein* in a feed, rather than the percentage of lysine in the feed, is the most important value to use

TABLE 1–3
EFFECT OF DIETARY LYSINE CONTENT ON GROWTH OF THE HORSE

	SBM[1]	CSM[1]	CSM + Lysine
Lysine (%)	0.65	0.49	0.65
Initial Wt (lbs)	580	580	580
Gain (lbs/day)	1.23	1.01	1.23
Feed Efficiency (lbs feed/lb of gain)	12.2	15.5	12.3

[1]Soybean meal (SBM) or cottonseed meal (CSM) used in the *ration.* All three rations contained the same total *protein* content.[55]

TABLE 1–4
LYSINE CONTENT OF FEEDSTUFFS[50]

Feed	% Protein in Air Dry Feed	% Lysine in Air Dry Feed	% Lysine in the Protein*
Alfalfa	13.5–19	0.6–0.9	5.0
Bran	15	0.5–0.6	3.8
Brewers Grains	24	0.9	3.7
Calf-Manna†	23	1.4	6.5
Cereal Grains	14–12.5	0.2–0.5	3.0
Cottonseed Meal	34–45	1.2	3.0
Fish Meal	59	4.8	8.0
Grass	5–12	0.2–0.4	3.8
Linseed Meal	34	1.2	3.4
Meat Meal	51	3.6	7.0
Soybean Meal	40–45	3.0	7.5
Start-To-Finish‡	27	1.9	7.0

*The most important value to consider when comparing the lysine content of different feeds [(% *lysine* ÷ % *protein*) × 100].

†Carnation-Albers, 6400 Glenwood, Box 2917, Shawnee Mission, KS 66201.

‡Milk Specialties Co., Box 278, Dundee, IL 60118.

when comparing different feeds with respect to their lysine content. As shown in Table 1–4, the percentage of lysine in the protein is high in alfalfa, soybean meal, and animal source feeds (such as fish meal, meat meal, and milk casein). These feeds will provide adequate lysine to meet the growing horse's needs; if they are not included in the *ration* it will frequently be deficient in lysine. These protein sources are also relatively high in other essential *amino acids* needed by the horse for growth.

Bacteria present in the *rumen* (or forestomach) of cattle and sheep and in the *cecum* of the horse are able to utilize *urea* or other substances containing nonprotein nitrogen (NPN) to synthesize *protein*. In cattle and sheep, as well as other *ruminants,* the protein produced by these bacteria passes to the stomach and small intestine, where it is digested and absorbed. Thus, a nonprotein source of nitrogen, such as urea, may be fed to ruminants to provide them with protein. However, feeding nonprotein nitrogen is of little value in providing protein to the horse. The cecum, where the bacteria convert nonprotein nitrogen to protein, is past the stomach and small intestine, where most protein is digested and absorbed (see Glossary Fig. 2). Because of this, much of the nonprotein nitrogen fed to the horse is absorbed from the small intestine and excreted in the urine before it reaches the cecum. That which reaches the cecum may be converted to protein, but little protein is utilized from the cecum or large intestine. Thus, although some of the nonprotein nitrogen fed to the horse may be converted to protein, much of that is excreted in the feces.

Excessive intake of nonprotein nitrogen is poisonous. Before bacteria use nonprotein nitrogen to synthesize *protein,* they convert it to ammonia. If an excessive amount of nonprotein nitrogen is ingested, toxic quantities of ammonia are absorbed. Initially, affected animals wander aimlessly and are incoordinated. Following this, they press their heads against fixed objects, go down, become comatose, convulse, and die. The horse can tolerate levels of nonprotein nitrogen, such as *urea,* several times greater than those that are toxic to cattle or sheep. Therefore, *rations* made for cattle or sheep that contain urea may be fed to the horse. Even though the urea is of little benefit to the horse, it is not harmful in the amounts present in these rations.

Adequate quantities of *protein* are necessary in the *ration* (Appendix Table 2). **Feed intake, growth, physical activity, physical endurance, and production of milk or fetal development are greatly impaired when protein intake is inadequate** (Fig. 1–3).

If more *protein* is ingested than needed, nitrogen, in the form of ammonia, is removed from the *amino acids* that make up the protein. The remainder of the amino acid is used for *energy,* or if energy is not

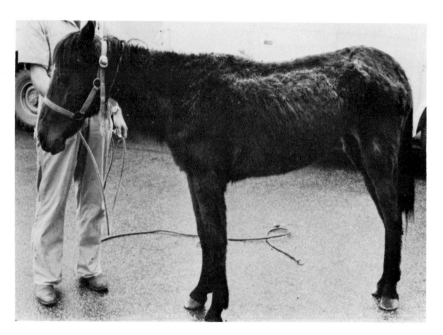

Fig. 1–3. Protein deficiency. Long, scruffy-looking hair coat, which occurred primarily as a result of a severe protein deficiency. A vitamin A deficiency may cause a similar effect. This horse was also suffering from an energy deficiency as a result of inadequate feed intake.

needed at that time, it is stored as *fat* or *glycogen* for later use as energy. The ammonia that is removed from the amino acids is converted by the liver to *urea*, which is then excreted in the urine by the kidney. This increases the ammonia smell in the urine and can be noticed in poorly ventilated stables when horses are fed higher protein-containing feeds such as *legume roughages*. It also adds nitrogen to the pasture, although this is a tremendously expensive way to fertilize.

The more *protein* eaten above the animal's needs, the more work the liver and kidney must do to change ammonia to *urea* and excrete it. It was believed at one time that this increased work load predisposed the animal to kidney disease. It has been well proven that this is not true. However, if an animal develops disease of the kidney or liver, he has a decreased ability to perform these functions, and the amount of ammonia and urea in his body increases. Decreasing protein intake is, therefore, an important part of the management of an animal with liver or kidney disease. Ammonia or urea at extremely high levels makes the animal sick, eventually resulting in *coma* and death. A greater amount of urine is excreted in an attempt by the body to flush out these waste products in a lower concentration, which requires less work by the diseased kidney. To compensate for the increased water loss in the urine, the animal must drink more water, so the earliest signs of kidney disease are increased water intake and urine excretion. But even when kidney disease is not present, protein intake greater than the animal's needs may increase the amount of urine that must be excreted to rid the body of excess ammonia and urea. Therefore, the results of excessive protein intake may resemble the early signs of kidney disease, leading to the mistaken belief that high protein intake causes kidney disease and lowered protein intake alleviates it. Liver and kidney disease may be caused by a number of *infectious organisms* and toxins, but not by the amount of protein intake.

The utilization of *protein* for *energy* produces three to six times more heat than the utilization of *carbohydrates* or *fats*.[69] This may be beneficial in a cold environment, but contributes to excessive sweating and heat exhaustion during physical activity, particularly in a warm environment. Occasionally, allergies to a specific protein in certain feedstuffs occur. This results in *urticaria* or hives (frequently called *protein bumps*, Fig. 1–4) over all or small portions of the animal's body, or less commonly, in diarrhea or respiratory problems. Caution must be used before attributing these signs to diet, since other factors may be responsible, such as insect bites, drugs, and insecticides. Excessive protein intake may also contribute to bone problems in the growing horse (see Chapter 9).

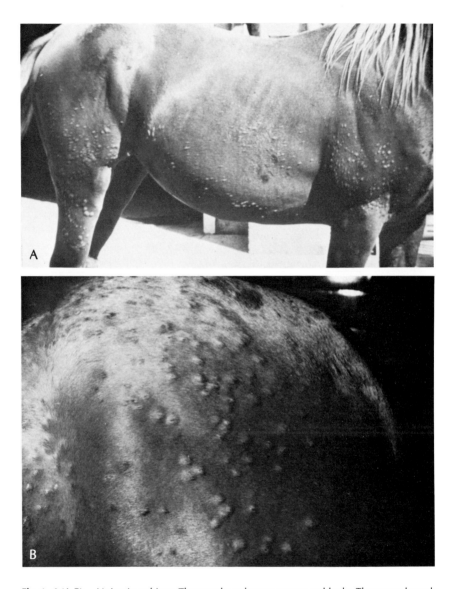

Fig. 1–4 (A,B). Urticaria or hives. These welts or bumps appear suddenly. There may be only a few or there may be many all over the head and body. They are caused by an allergic response to things such as feed, insect bites, drugs, and insecticides. When caused by feed, they may be referred to as protein bumps.

MINERALS

The major *minerals* of concern in feeding the horse are calcium, phosphorus, and salt (sodium chloride). A number of mineral deficiencies and toxicities may occur in the horse, but with the exception of those of selenium, calcium, and phosphorus, they are uncommon or rare when average or better quality feeds are fed, *trace-mineralized* salt is available, and excess mineral-containing *supplements* are not added to the *ration*.

Calcium and Phosphorus

Calcium and phosphorus comprise about 70% of the *mineral* content of the animal's body and from one third to one half of the minerals in milk. About 99% of the calcium and over 80% of the phosphorus are in the bones and teeth. **Horses are more likely to suffer from a lack of calcium and phosphorus than from any other minerals.** The amounts needed in the horse's ration are given in the Appendix Table 2. The diet must not only contain adequate amounts of calcium and phosphorus, but the intestine must be able to absorb them. Phosphorus bound to an organic substance such as that present in *phytate*, is less available for absorption than that present in *inorganic* minerals.[62,63] *Cereal grains* are high in phytate and the content increases with maturity.[28] Thus, the phosphorus present in cereal grains is less available than that present in *roughages* or minerals. The true digestibility of the phosphorus present in most *concentrates* is 29 to 32%; in roughages, 44 to 46%; and in inorganic minerals, 58%.[62,63]

The calcium and phosphorus requirements given in Appendix Table 2 assumes that 55% of the calcium and 35% of the phosphorus will be absorbed. Since the absorption of the calcium and phosphorus from the horse's *ration* is generally higher than this, absorption does not usually need to be considered in ensuring that the horse receives adequate calcium and phosphorus. However, several factors may decrease the efficiency of absorption.

Phytate (as well as excess phosphorus in any form) and *oxalates* bind other *minerals (cations)*, such as calcium, decreasing their absorption.[28,62,63] Thus, even if the amount of calcium present in the *ration* meets the animal's dietary requirements, adequate quantities are not absorbed and calcium deficiency may occur. This effect of excess dietary phosphorus becomes more important when the calcium content of the diet is low. In one study with 2-year-old ponies, net skeletal calcium deposition decreased by more than 50% in those ponies fed a diet that barely met their calcium requirements (0.35%) but contained excessive phosphorus (1.2%), as compared to another group of ponies fed a diet that contained the minimum amount of both calcium and phosphorus needed to meet their requirements (0.4% Ca and 0.2% P).[63] Excessive amounts of oxalates may be present in some plants, e.g., Sertaria sphacelata, Panicum (giant, blue or green panic grass), Paspalum spp., Sporobolus spp., Cenchrus ciliaris (buffel grass), halogeton, greasewood, beets, docks, and rhubarb. If the horse consistently eats these plants over an extended period, calcium deficiency may result. In addition, insoluble oxalate crystals are deposited in the kidneys, resulting in kidney damage.

Calcium deficiency results in the mobilization of calcium from the bone, which is replaced by a proliferation of fibrous connective tissue. This increases the size of the bone at that location. In mature horses this is most noticeable in bones of the skull and jaw, resulting in the condition commonly called "*big head*" (Fig. 1–5). "Big head" was seen more in the past in horses on diets consisting largely of bran, and therefore is also referred to as "bran disease." Bran is high in phosphorus and low in calcium (Appendix Table 1). Calcium deficiency in the growing horse most commonly causes leg problems, such as enlarged joints, *splints,* and the *epiphysitis* syndrome (see Chapter 9). Lameness may occur prior to the appearance of these signs. Further, the lameness may shift from one leg to another. A severely affected horse may have a gait similar to that of a hopping rabbit.[41] A "cardboard" sound may be heard on percussion of the sinuses.

Excessive calcium in the diet has less effect on phosphorus absorption than excessive phosphorus has on calcium absorption.[62,63] In ponies, high dietary calcium intake decreased phosphorus absorption, although the decrease was small.[61] Thus, excessive calcium in the diet is much less detrimental than excessive phosphorus. This is illustrated by the relationship between the amount of calcium to phosphorus in the *ration,* called the *calcium-phosphorus ratio (Ca : P).* If quantities of both calcium and phosphorus in the ration are adequate to meet the animal's requirements, the Ca : P ratio in the ration of the mature animal can vary from 0.8 : 1 to 8 : 1, and in the growing horse from 0.8 : 1 to 3 : 1, without resulting in problems. As an additional safeguard, and because of variation in the calcium and phosphorus

$\longrightarrow$

Fig. 1–5 (A,B).   Calcium deficiency, or phosphorus excess, which decreases intestinal calcium absorption, decreases the blood plasma calcium concentration. This stimulates excessive secretion of parathyroid hormone, causing a condition referred to as nutritional secondary hyperparathyroidism. Parathyroid hormone mobilizes calcium from the bone to compensate for the dietary calcium deficiency. The minerals mobilized from the bone are replaced by fibrous tissue, which increases the size of the bone. In the mature horse, as shown in A, it is most noticeable on each side of the nose one-half way between the eyes and the nostrils, giving rise to the common name for this condition "big-head." The skull of an affected horse (B) shows the enlargements on each side of the nose and on the lower jaws or mandible. (Courtesy of J. R. Joyce, Texas A&M University.)

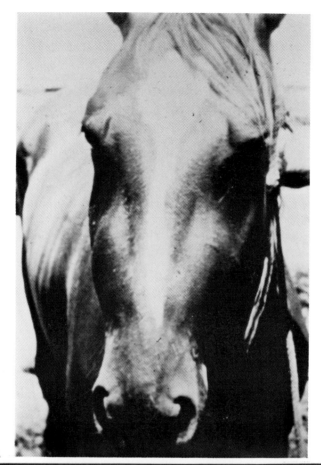

A

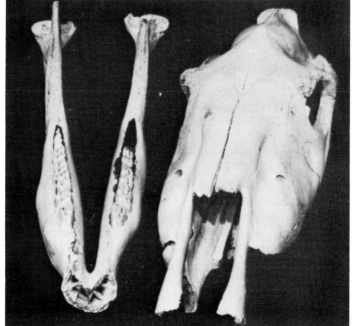

B

15

content of feeds, a ratio of less than 1:1 is not recommended for any horse, for the mature horse, a ratio of greater than 6:1 is not recommended, and for the growing horse, a ratio not greater than 2:1 is safest. If the amount of calcium or phosphorus in the total ration is less than that necessary to meet the growing horse's requirements, or if the amount of one *mineral* with respect to the other is outside of these ratios, *epiphysitis* and improper bone growth may occur.[62]

Excessive calcium can decrease the absorption of a number of *trace minerals* (such as zinc, manganese, iron), and may cause a deficiency if the *ration* contains only marginal amounts of these minerals. This is an uncommon problem. Excessive zinc, manganese, iron, and other minerals decrease calcium and phosphorus absorption. Methods for determining an excess or deficiency of calcium or phosphorus in the ration are described in Chapter 3.

Many commercial *vitamin-mineral supplements* contain inadequate amounts of calcium and phosphorus and frequently contain a multitude of other ingredients which are of no known benefit to the horse. Calcium and phosphorus are the major supplemental minerals needed by the horse. They can be supplemented least expensively by using the minerals given in Appendix Table 5. The calcium and phosphorus in all of these minerals are readily available to the horse. There is little difference in the availability of calcium and phosphorus among the different mineral supplements. Mineral supplements similar to these can be obtained at most feed stores. However, they may be called by different names and may vary in their mineral content.

Salt

Salt should always be available free-choice for all horses. The mature horse, with minimal use, will consume, on the average, about ½ lb (0.2 kg) of salt per week, although this is variable. If salt is available, the horse will always consume enough of it to meet its needs. These needs can be met when either block or loose salt is provided, although consumption of the loose form is generally greater. The use of loose salt may be beneficial to increase the consumption of substances such as calcium and phosphorus added to the salt. There is little danger of excessive salt intake if palatable, nonsaline water is available. A lack of salt results in a decreased and abnormal appetite, which in turn, results in weight loss and eating or licking of substances such as dirt, rocks, wood, and urine.

Trace-mineralized salt is recommended to assist in providing some of the *microminerals* occasionally needed. The additional cost of trace-mineralized salt is small and can be considered low-cost insurance against a deficiency in these microminerals. The minerals such a product generally contains are salt or sodium chloride (98%), zinc

Fig. 1–6. Plain salt, iodized salt, and trace-mineralized salt. Trace-mineralized salt, available for free-choice consumption, is recommended.

TABLE 1–5
MINERALS NEEDED OR TOXIC TO THE HORSE AS COMPARED TO THOSE PRESENT IN FEEDS[50]

Mineral	Amount in Total Ration		Amount Present in		
	Needed	Toxic	Alfalfa	Grass	Grain
Sodium, %	0.35		0.15–0.20	0.02–0.4	0.01–0.2
Potassium, %	0.5		1.5–2.5	1.5–2.5	0.3–0.5
Magnesium, %	0.1		0.25–0.40	0.15–0.6	0.1–0.2
Sulfur, %	0.15		0.2–0.5	0.15–0.25	0.15–0.4
Iron, ppm	50		170–400	150–250	30–90
Zinc, ppm	<40*	200†	17–22	17–22	17–50
Manganese, ppm	<40*	—‡	25–30	40–190	6–45
Copper, ppm	9	—‡	9–15	5–25	4–9
Cobalt, ppm	0.1				
Selenium, ppm	0.1	5.0			
Iodine, ppm	0.1	4.8			

*The amount required is not known but is something less than these amounts.
†Although levels of 9000 ppm of zinc may be required to cause zinc toxicity,[50] levels of 200 ppm may cause a calcium deficiency, epiphysitis, and lameness in the growing horse.
‡Nutrients known to be toxic to other species, but levels at which they are toxic to horses are not known.

(0.35%), manganese (0.28%), iron (0.175%), copper (0.035%), cobalt (0.007%), and iodine (0.007%). Iodized salt also contains 0.007% iodine. Plain salt does not contain any iodine or other minerals. Trace-mineralized salt is generally a blue-gray or dark reddish-brown color; iodized salt, a lighter red; and plain salt, white (Fig. 1–6). However, the colors may vary, so don't rely on them. Look at the labels. The amount of minerals needed by the horse and the levels that are toxic to it, as compared with the amounts present in natural feeds, are shown in Table 1–5.

Selenium

Selenium deficiencies and toxicities both may occur in the horse. Soil (and therefore plant) selenium content varies greatly. Most of the area around the Great Lakes and the eastern and northwestern United States are low in selenium (Fig. 1–7). *Forages* and *grains* grown in these areas may contain less than 0.1 ppm of selenium. Less than this amount of selenium may result in a deficiency. In contrast, some areas

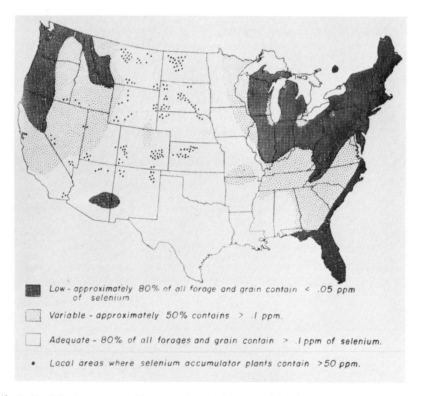

Fig. 1–7. Selenium content of forages and grains by regional distribution in the United States. (From Kubota, J., and Allaway, W. H.: Journal of Dairy Science, *58*(10):1563, 1975.)

in the Rocky Mountain—Great Plains regions have soil selenium contents high enough (greater than 0.5 ppm) to result in toxicity from ingestion of cultivated crops and grasses grown in these areas.

A clinically apparent selenium deficiency occurs primarily in foals 1 to 10 days of age, but may occur up to 8 months of age, and rarely affects mature horses. Mild deficiencies may decrease the animal's immune response to *infectious* diseases and decrease growth rate. More severe deficiencies cause muscle damage resulting in stiffness, lameness, listlessness, muscle pain, and an increased release of enzymes from the muscle into the blood. The foal may go down and die within a few hours, but generally lives for one to two days. Death is due to respiratory failure, because of the effect of the deficiency on the muscles of respiration. Sudden stress or muscular exertion may initiate the onset of these symptoms. This condition is often called "white muscle disease" because of the white striations and patches that occur in the muscles, particularly those of the hind legs and neck (Fig. 1–8). Although much less common, selenium deficiency may occur in the mature horse, affecting muscles used for chewing or those of the legs, resulting in a stiff gait. The muscle cell damage may result in the release of enzymes, such as SGOT, resulting in an increase in the *plasma* concentrations of these enzymes. If cellular damage is

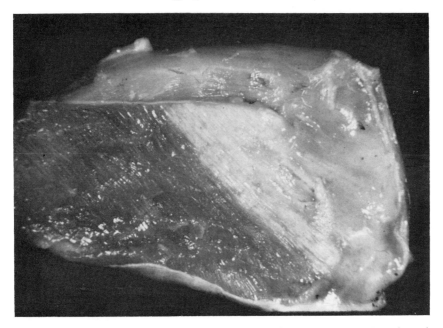

Fig. 1–8. White muscle disease induced by selenium deficiency. A cross section of muscle from an animal that died from this disease showing the white areas of ischemic muscle necrosis, which gives this disease its name.

severe, myoglobin may be released from the muscles and excreted in the urine. Myoglobin gives the urine a coffee-colored appearance.

Selenium deficiency in cows increases the incidence of retained membranes (afterbirth or *placenta*) following calving. Although selenium deficiency has not been shown to have this effect in the mare, it should be considered if this problem occurs. A selenium deficiency has been associated with a higher incidence of reproductive-related diseases in the mare, including *uterine* infections, repeat breeding, early embryonic deaths, and abortions.[80,81]

Because selenium deficiency resembles *azoturia* and *tying-up* (see Chapter 5), and because selenium stabilizes cell membranes, selenium injections or selenium *supplements* in the *ration* may be given to horses who are repeatedly affected by the syndrome. This appears to be beneficial in some cases (see Vitamin E later in this chapter). Selenium administration may also enhance physical performance of horses deficient in selenium, as indicated by blood or *plasma* selenium concentrations of less than 0.06 to 0.1 ppm, but showing no clinical signs of the deficiency.[82] It is emphasized however, that selenium administration has no beneficial effect on horses that are not selenium deficient, and if given in excess, may be harmful.

Selenium deficiencies may be treated or prevented by giving a selenium injection into the muscle (1 ml of E-Se* per 100 lbs body weight or 2 ml/100 kg), feeding a *ration* containing 0.1 ppm selenium in its *dry matter*, or allowing free access to salt containing 20 to 40 ppm selenium. *Trace-mineralized* salt containing selenium is available for sheep and cattle and may be used for the horse who is repeatedly affected by *azoturia* or *tying-up*, or in selenium-deficient areas. Levels of greater than 2 ppm should not be added to the ration, since excess selenium is toxic.

If selenium deficiency occurs in foals, in addition to feeding a selenium-containing salt or adding selenium to the pregnant mare's *ration*, the mare should be given a selenium injection three weeks to three months prior to foaling. Selenium injections early in pregnancy may be teratogenic in sows. Although this hasn't been shown to occur in mares, as a safety precaution selenium injections should not be given early in pregnancy. In addition to giving a selenium injection to the mare, injections should be given to the foal because only a limited amount of selenium crosses the *placenta* or is secreted in the milk.[80,83] In areas where selenium deficiency problems have been encountered, the foal should be given a selenium injection at birth and again at 1, 3, and 6 months of age.

Selenium concentrations of less than 0.06 ppm in the blood or *plasma*, 0.20 ppm in hair or hoof, and 0.17 ppm in the liver (wet

*Burns Biotec Labs, Oakland, CA 94621.

weight) are indicative of a deficiency. Toxicity is indicated by levels greater than 0.3 ppm in the blood or plasma, 1.2 ppm in the liver (wet weight), and 5 to 20 ppm in the hoof or hair. A selenium deficiency is also indicated by the activity of the selenium-containing enzyme glutathione peroxidase of less than 5 to 8 units; normal being greater than 15 to 30 units.[80,84]

Selenium toxicity may occur as a result of accumulation by certain plants, even in areas where soil selenium levels are relatively low. Greater than 5 ppm selenium in the *ration* is toxic. Plants such as milk vetches (Astragaluses), woody aster (Aster xylorrhiza), golden weed (Oonopsis), and prince's plume (Stanleya) may accumulate selenium at levels as high as 10,000 ppm, even on soils containing only moderate amounts of selenium. These plants are called selenium indicator plants because they can grow on high selenium soils on which many other plants will not survive. Selenium content is highest during the growth period of the plant. When the selenium content is high, it gives these plants an unpleasant garlic-sulfur odor which makes them relatively unpalatable. The odor is increased by rubbing the leaves together. Horses will avoid eating these plants if other feed is available, so pastures containing these plants may be used safely if they are not overgrazed.

Plants that most often cause selenium toxicity are those that do not require selenium for growth but will accumulate up to several hundred ppm of selenium when soil selenium levels are high. These include asters (or Machaeranthera), four-winged saltbush (Atriplex), wheatgrass or bluestem (Agropyron), gumweed (Grindelia squarrosa), broomweed, snake weed, or match weed (Gutierrezia sarothral), Sideranthus, Comandra, and Castilleja. When soil selenium is not excessive, some of these plants are valuable *forage* plants. Many cultivated crops, plants, *grains*, and native grasses may contain 1 to 30 ppm selenium when soil selenium levels are high. Selenium is taken up by plants more readily from alkaline soils. Early Great Plains and Rocky Mountain settlers referred to selenium poisoning as "alkali disease," believing the disease was caused by the high salt content of the water in these semiarid regions.

Selenium toxicity is more common than deficiency in the mature horse. Intake of high levels of selenium may result in blind, aimless wandering, circling, stumbling, and incoordination. The eyelids may be swollen and the *corneas* cloudy. There may be bloody froth from the nose, abdominal pain, and death due to respiratory failure. Much more commonly, however, lower amounts of selenium are ingested and cause listlessness, weight loss, *anemia, emaciation,* and a rough, coarse, dry, brittle hair coat. Hair may be lost from the mane and tail switch, giving rise to the common name, "bob-tail disease," for this condition (Fig. 1–9).

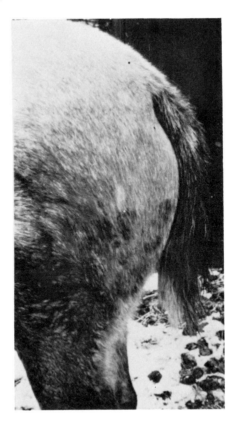

Fig. 1–9. "Bob-tail disease." The condition takes its name from a loss of hair from the tail induced by selenium toxicity. This horse had hair selenium levels of 17 ppm; greater than 5 ppm indicates toxicity.

Animals affected by selenium toxicity walk stiff-legged with tenderness followed by pronounced lameness. In mild cases, a ring of abnormal hoof growth may occur. In more severe cases, transverse breaks and cracks develop in the hoof wall (Fig. 1–10). When new hoof growth occurs, these breaks move downward and the old hoof separates from the new growth. The old hoof may not be completely shed, resulting in ragged, long hoofs, turned up at the ends.

Affected animals should be removed from selenium-containing *forage* or *grain,* and their feed supplemented with low selenium-containing grain and a high-*protein ration* (25%). Providing feeds low in selenium, when horses are grazing pastures containing plants high in selenium, assists in prevention.

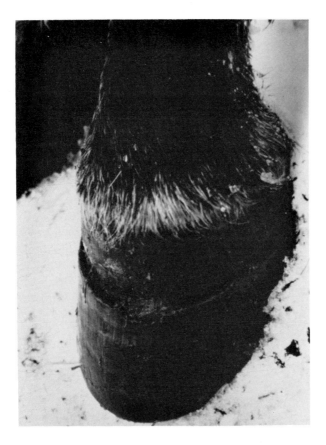

Fig. 1–10. Break in the hoof wall induced by selenium toxicity. This horse had selenium concentrations of 25 ppm in the hair and 11 ppm in the hoof wall; greater than 5 ppm in either indicates toxicity.

Iodine

The amount of iodine present in *trace-mineralized* salt and iodized salt will supply all of the iodine needed by the horse. Since less than 5 ppm of iodine in the horse's total *ration* will cause toxicity,[50] care should be taken not to add excessive quantities of iodine to the ration. Some seaweed- or kelp-containing products may contain high levels of iodine.[3,11,12] Iodine is also present in many commercial *vitamin-mineral supplements* which if fed in amounts greater than those recommended by the manufacturer, may cause iodine toxicity. Iodine toxicity may result in *chronic* respiratory disease that does not respond to treatment for an *infectious* condition. It also causes *goiter*.

Goiter is an enlargement of the thyroid gland that may be caused by either an excessive or an inadequate iodine intake. The thyroid gland consists of two lobes, one located on each side of the windpipe (trachea) just behind the throatlatch (Fig. 1–11). Goiter occurs most commonly in foals; the foals may be weak at birth and die within the first few days of life.[11] The thyroid gland will be significantly larger than the normal of 15 g, and its iodine content will be altered from the normal *dry matter* content of 0.2 to 0.5%. With iodine toxicity, the iodine content may be increased to as high as 1%, and with iodine deficiency, it is below normal. An iodine deficiency or toxicity may be present in foals at birth if the mare received less than 1 to 2 mg or greater than 40 mg of iodine daily during pregnancy. Iodine is readily secreted in the milk, so that if the lactating mare is consuming excessive iodine, the nursing foal is affected. Excessive iodine intake may be confirmed not only by determining total iodine intake, but also by determining the mare's blood *plasma protein*-bound iodine concentration, which is increased above the normal of 16 to 27 μg/l.[11] However, the foal may be affected by iodine toxicity even when the mare's plasma protein-bound iodine level is normal.[11]

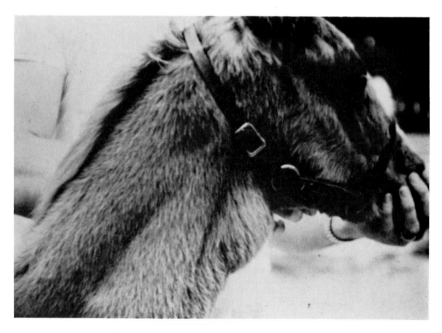

Fig. 1–11. Goiter. Enlarged thyroid glands, which may occur as a result of either a deficiency or excess of iodine. Goiter may also be caused by thiouracil-, thiourea- and methimazole-containing drugs, or by the ingestion of large amounts of feeds high in perchlorates, nitrates, and thiocyanates, such as kale, white clovers, rutabaga, and turnips. These drugs and plants contain substances that decrease the secretion of thyroid hormone.

Zinc

A zinc deficiency has never been reported in horses on natural feeds. An experimental deficiency has been induced, but only when the zinc content of the *ration* was reduced to 4 ppm.[29] Natural feeds contain 4 to 12 times this amount. Excess zinc in the diet decreases the absorption of both copper and calcium and may cause a calcium deficiency. This may occur as a result of a ration containing as low as 200 ppm zinc or of water containing in excess of 15 ppm zinc (Table 1–1). Excess zinc in the ration may be caused by excess supplementation or by pastures contaminated with efflux from nearby smelters. Excess zinc in the water may occur when galvanized pipes are connected to copper. This results in *electrolysis*, releasing zinc from the galvanized pipes into the water. Excess zinc intake by the growing horse may result in *epiphysitis*, lameness, bog spavins (increased tibiotarsal effusions), *osteochondritis dissecans*, and an increase in the *renal clearance ratio* of phosphorus[85] (see Chapter 9). A zinc excess can be confirmed by measuring *plasma* or serum zinc concentration of greater than 2 ppm. The blood for a plasma or serum zinc analysis must be taken with a glass syringe or into a glass vacuum tube, and must not be allowed to touch a cork or rubber stopper. If the blood is put in a tube, the stopper should be covered with waxed paper or plastic.

Copper, Molybdenum, and Sulfur

Copper deficiencies and toxicities are uncommon in animals other than ruminants. Since the mineral concentrations in feed and water that might cause copper imbalances in the horse are not known, the values given are for ruminants.[86] Horses are much less susceptible to either copper deficiency or excess than are ruminants, therefore, dietary mineral concentrations greater or less than those given would be necessary to cause dietary copper imbalance in the horse.

In *ruminants*, copper deficiencies may occur when the copper content of the *ration* is less than 11 to 13 ppm and there is less than 2 parts copper to 1 part molybdenum, i.e., a Cu:Mo ratio of less than 2:1. When the *ration* contains greater than 13 ppm copper, the amount of molybdenum is not important, but below this level, molybdenum at a concentration of greater than one half that of copper may cause a copper deficiency. Copper toxicity may occur in ruminants when the ration contains over 15 times more copper than molybdenum. Copper deficiency or toxicity is the only effect of an excess or deficiency of molybdenum in the ration. Horses have been reported to tolerate dietary copper levels as high as 790 ppm.[88]

Copper deficiency may also occur as a result of excess sulfur or sulfate in the diet. Sulfur concentrations in water and feed may be reported as sulfate (SO_4), sulfur, or sulfate sulfur. To convert the amount of sulfur or sulfate sulfur to sulfate, multiply by three. More than 3000 ppm sulfate in the ration may cause a copper deficiency in ruminants receiving a ration containing less than 8 to 11 ppm copper. This would occur only if excess sulfur-containing minerals were added to the ration. The upper safe level of sulfate recommended in the water is 250 ppm (Table 1–1). However, a sulfate concentration in water of 2700 ppm has been reported to have had no effect on 660-lb (300 kg) heifers over a three month period, whereas levels of 3400 ppm caused weight loss. Sulfur deficiencies are unlikely to occur if the horse is receiving adequate *protein,* since sulfur is present in protein. Copper absorption by the horse is also reduced by excess zinc in the ration.[87]

Copper deficiency may cause stillbirth, *anemia,* a *chronic* diarrhea called "teart," loss of hair pigmentation, and in the young growing lamb, calf, or pig, a paralysis of the back legs called "swayback" or "enzootic ataxia." Copper deficiency in newborn or growing *ruminants* may also result in lameness, stiffness, spontaneous fractures, erosion of articular cartilages, enlarged joints, *epiphysitis,* or contracted flexor tendons. This has also been reported in foals which were thought to have an inability to properly utilize a dietary copper intake normally adequate for the horse.[89,90] During growth, a horse with epiphysitis and contracted flexor tendons was reported to respond to copper supplementation and the signs reoccur when supplementations was stopped.[90] However, a copper deficiency is not responsible for the vast majority of these cases in growing horses, and therefore, copper supplementation is unlikely to be of benefit in most cases. Copper deficiency in the mature ruminant may result in rupture of an *artery* or of the aorta and fatal hemorrhaging. A decrease in the *plasma* copper concentration in aged mares has been related to increased frequency of rupture of the *uterine* artery during pregnancy.[91] In lambs, copper toxicity results in depression, weakness, *acute hemolytic anemia, icterus, hemoglobin* excretion in the urine, often excessive thirst, and death within a few hours to a few days. It is not known to occur in horses.

Dietary copper imbalances in the horse are rare, but may be diagnosed by finding copper concentrations significantly outside the normal of 0.7 to 2 ppm in the *plasma* or 3 to 15 ppm in the hoof. A deficiency may also be indicated by a hair copper content of less than 8.5 ppm.

Potassium, Manganese, and Magnesium

Potassium deficiencies occur with excessive sweating, anorexia, and diarrhea, and result in muscle weakness, lethargy, and decreased feed intake (see Chapter 5). Excess potassium ingestion is not a problem so long as water is available and urine excretion is normal. Adding several ounces of potassium chloride to the ration daily may in some instances decrease wood chewing, even though there is adequate potassium in the ration. However, in the majority of cases, wood chewing is not decreased by potassium chloride supplementation.

Manganese deficiency has never been reported in the horse. Deficient manganese in species other than the horse may cause any of the following effects: delayed estrus, decreased conception, poor growth, abortion or stillbirths, but more often the birth of weak young with limb deformities such as enlarged joints, knuckled-over pasterns, twisted forelimbs, and thickened, well calcified, brittle, shortened bones resulting in lameness, stiffness, joint pain, and a reluctance to move.[87] A manganese concentration of less than 0.02 ppm in plasma or blood, 6 ppm in liver dry matter, or 4.5 ppm in kidney *dry matter* is indicative of a manganese deficiency in *ruminants*.[87] Manganese toxicity does not occur naturally, even with the ingestion of large amounts over a long period of time.

A decrease in the *plasma* magnesium concentration may occur in lactating cows, and rarely in other cattle feeding on lush, green growing pasture in the spring. Although this condition, called grass *tetany*, has been reported in the horse, it is rare. It results in muscle tetany. Death occurs within a few hours if magnesium-containing solutions are not given. Giving excessive magnesium or giving it too rapidly into the *vein* will cause sudden death. In areas where grass tetany occurs, its occurrence may be decreased in both horses and cattle by feeding 1 to 2 ounces (30 to 60 g) of magnesium oxide daily. Magnesium sulfate (epsom salts) is not recommended because it is much lower in both magnesium content and *palatability*, although magnesium oxide is also fairly unpalatable. Although for cattle it is best to add magnesium oxide to a *grain*, or a mixture of 45% salt, 45% magnesium oxide, and 10% soybean meal, adding as little as 5% magnesium oxide to the salt may be effective in preventing grass tetany in horses. Magnesium intake should be increased several weeks before and continued throughout the time of year when grass tetany is expected to occur. The only problem likely to occur as a result of excess magnesium ingestion is loose stools.

Iron and Cobalt

Iron deficiencies occur only if there is a loss of blood (see Chapter 5). Blood loss may be inapparent, such as that caused by lice and intestinal parasites *(worms)*. The major effect of an iron deficiency is *anemia*, which is treated by giving an iron injection. Iron dextran, although commonly given to other species, should not be given to the horse. Intramuscular injection of iron dextran has reportedly resulted in the death of three horses.[92] Although the ingestion of excess iron decreases phosphorus absorption, excess intake of the magnitude required to cause a phosphorus deficiency does not occur unless large amounts are added to the *ration* (for swine, more than 1500 ppm).

Cobalt is needed by intestinal bacteria to synthesize *vitamin* B_{12}; therefore, a cobalt deficiency results in a vitamin B_{12} deficiency. A deficiency in cobalt or vitamin B_{12} in the horse is rare, if it occurs at all. Horses have remained in good health while grazing pastures so low in cobalt that *ruminants* confined to these pastures died.[16]

VITAMINS

Vitamin A is the only vitamin that may be inadequate in rations routinely fed to the horse. The precursor of *vitamin A, beta-carotene,* is present in green *forages* at concentrations many times over the horse's requirements. Its content in the forage decreases with maturity and with storage. An approximation of the amount of beta-carotene present in a forage can be judged by the amount of green color; unless the forage is brown or yellow, it will contain more than enough beta-carotene to meet the horse's vitamin A requirements. In addition, the liver stores sufficient vitamin A to supply the animal's requirements for three to six months. Because of this, if the horse consumes fresh green forage for a period for four to six weeks, it will have sufficient vitamin A to meet its needs for three to six months.[50] If poor quality forages are being fed for long periods, 25,000 IU per day of vitamin A should be added to the *ration* for mature horses. During the last 90 days of pregnancy, during lactation, and for the weanling twice this amount is recommended. For the yearling, 75,000 IU per day is recommended. Feeding less than one half or 5-fold greater than these amounts may decrease growth rate, *hematocrit,* and *plasma* concentrations of iron, albumin, and cholesterol, even though the classic signs of vitamin A deficiency or toxicity as described further on may not occur until vitamin A levels either much lower or as much as 50-fold greater than that recommended are fed.[93] Instead of adding vitamin A to the ration, it may be injected into the muscle or under the skin. Use a water-soluble emulsion of vitamin A at a dosage of 3000 IU per lb of body weight (6660 IU/kg). This amount will completely saturate the liver's storage capacity and, therefore, does not need to be repeated for at least three months. Injecting four times this amount may cause vitamin A toxicity.

Many of the symptoms of *vitamin A* deficiency and toxicity are similar. The vitamin-A-deficient animal is more susceptible to pneumonia, diarrhea, and *infectious* conditions of the vagina and *uterus,* and has a decrease in growth and reproductive ability. Exces-

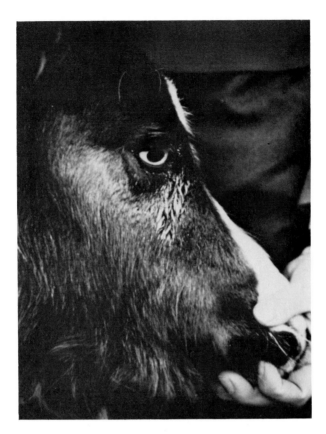

Fig. 1-12. Excessive tearing caused by vitamin A deficiency. (From Evans, J.W., Borton, A., Hintz, H. F., and Van Vleck, L.D.: The Horse. W.H. Freeman and Company. Copyright © 1977.)

sive tearing, night blindness, and an increased sensitivity to light (photophobia) are suggestive of a vitamin A deficiency (Fig. 1–12). The *cornea* may become inflamed and ulcerated. The hair becomes rough, dry, dull, and brittle, and shedding is prolonged, resulting in a long, scruffy looking hair coat. These same hair coat changes may occur with a *protein* deficiency (Fig. 1–3). Vitamin A deficiency or toxicity also impairs the normal breakdown and deposition of minerals into bone. Vitamin A deficiency may be confirmed by finding *plasma* vitamin A levels of less than 20 μg%, although signs generally are not evident until it falls to less than 10 μg%.

Vitamin D_2 is present in ample quantities in good quality sun-cured feeds. In addition, a substance (7-dehydrocholesterol) synthesized in the body in ample amounts is converted in the skin by the ultraviolet rays from the sun to vitamin D_3. Vitamins D_2 and D_3 are both utilized by the horse. If the animal receives either sun-cured feeds **or** is outside a few hours a day (even if it is cloudy), a vitamin D deficiency will not occur. As a result, vitamin D deficiency in the horse is a rare occurrence. However, horses that are continually stabled during the day may be supplemented with a maximum of 3 IU/lb of body weight (1.4 IU/kg) daily.

Vitamin D$_2$ and D$_3$ must both be converted to another form in the liver (25-hydroxycholecalciferol) and finally changed to the active form in the kidney (1,25-dihydroxycholecalciferol). The active form of the vitamin is necessary for intestinal calcium absorption and bone formation. Vitamin D deficiency, therefore, results in an inability to absorb and properly utilize calcium. Bone demineralization and the disease called rickets occurs, resulting in enlargement of joints, bowing of the long bones, going down in the pastern, and decreased growth. However, vitamin D toxicity from adding excessive quantities to the ration is much more common than deficiency.

Toxicity occurs as a result of feeding *vitamin D* for several months at levels greater than ten times the animal's requirements of 3 IU/lb of body weight (6.6 IU/kg). Vitamin D toxicity occurs in a shorter period of time if a correspondingly greater amount is fed. Excess vitamin D causes an increase in the *plasma* calcium concentration and the deposition of bone in soft tissue where it does not belong. This often results in damage and functional impairment of these tissues, which can be serious in organs such as the heart, blood vessels, and kidneys. Excessive bone development may also cause joint stiffness and calcium enlargements on the bones.

Vitamin K$_1$ (phytylmenaquinone, or formerly, phylloquinone) is present in high amounts in green *roughages*, either fresh or dried. It is the biologically active form. Other forms of *vitamin K* are converted to it, primarily in the liver. *Vitamin K$_2$* (multiprenylmenaquinone, or formerly, farnoquinone) is also produced by bacteria in the horse's *cecum* and intestinal tract in ample quantities to meet the horse's requirements. There is also a commercially available synthetic vitamin K called vitamin K$_3$ (menaquinone, or formerly, menadione) which, like vitamin K$_2$, must be converted to the active form of the vitamin in the liver. Because vitamin K is necessary for blood coagulation, it is frequently given to *bleeders* (see Chapter 4). However, nasal hemorrhage is not due to a vitamin K deficiency and giving vitamin K is of no benefit in treating or preventing the condition.

Vitamin K deficiency occurs in the horse only as a result of ingestion of vitamin K antagonists. The only one of these likely to be ingested by the horse is dicoumarol. Dicoumarol is produced by a mold that occurs most commonly on sweet clover hay. Dicoumarol is not present in sweet clover pastures or hay not containing mold. Generally, dicoumarol-containing sweet clover hay must be ingested for several weeks to cause a vitamin K deficiency. Since vitamin K is necessary for blood coagulation, a deficiency results in excessive bleeding if the animal is cut. For this reason, it is safest to not have the horse eat sweet clover for at least a week before elective surgery is performed.

Vitamin K deficiency may also result in bloody diarrhea, hematomas, bleeding into the eyeball, abdomen, or muscle, and a bloody nose *(epistaxis)*. If excessive amounts of blood are lost, *anemia*, depression, weakness, a rapid, irregular heart rate, and death may occur. If blood loss is severe and the horse is in danger of dying, a blood transfusion is necessary to immediately restore blood clotting and blood volume. However, the administration of a stable emulsion of vitamin K_1* into the vein is effective within four to six hours.

Vitamin K toxicity occurs only as a result of the excessive administration of vitamin K, and causes rupture of the *red blood cells.*

Vitamin E deficiency has never been reported in the horse and excess vitamin E is not toxic. Vitamin E, also called tocopherol, is present in more than ample amounts in natural feedstuffs to meet the horse's requirements. *Roughages, cereal grains,* and particularly cereal germ *oils* and vegetable oils are high in vitamin E. Wheat germ oil is commonly added to the horse's *ration* to provide vitamin E.

The following effects of *vitamin E* deficiency have been noted in swine, dogs, chickens, and some laboratory animals. Vitamin E deficiency in males has caused degenerative changes in the testes and subsequent sterility. In females, it has caused failure of gestation and implantation, and increased embryonic, fetal, and neonatal deaths. However, a vitamin E deficiency has never been associated with reproductive problems in the horse or species of animals other than those just listed. As stated by the National Research Council, "there is no evidence to indicate that dietary supplementation of vitamin E helps resolve reproductive problems or in any way affects reproductive ability in horses."[50]

Vitamin E deficiency has been thought to be responsible for causing muscle damage, particularly in the foal. However, this is apparently due to selenium, rather than a vitamin E deficiency since *plasma* vitamin E levels are not decreased (see the discussion of selenium earlier in this Chapter).

Vitamin E and selenium are commonly given in the treatment of *exertion myopathy, tying-up syndrome,* and muscle inflammation and damage (see Chapter 5). One milliliter (cc) of E-Se† per 100 lbs body weight (2 ml/100 kg) is injected into the muscle. This may be repeated at weekly intervals for as many as four injections. Both vitamin E and

---

*Such as Synkavite (Hoffman-LaRoche).
† Burns Biotec Labs, Oakland, CA 94621.

selenium protect and stabilize muscle and other cell membranes and may increase blood circulation to the legs. In controlled studies in elderly humans with decreased circulation to the extremities, giving *vitamin* E was shown to markedly improve blood flow and walking ability.[94] Because of these effects, vitamin E and selenium may be beneficial in the treatment and prevention of these conditions in the horse, even though a deficiency of neither one is thought to be associated with the disorders.

In studies in several different species of animals, including the horse, high levels of *vitamin E* have been shown to increase the animals' immunity to *infectious* diseases. The practical application of this information, however, has not been demonstrated. High levels of vitamin E have been claimed to be of benefit during physical activity; however, there is no convincing evidence to support this contention. There appears to be no indication for or benefit from supplementing the horse's *ration* with vitamin E or substances high in vitamin E, such as wheat germ oil.

There are many **B-vitamins.** These are thiamine (B_1), riboflavin (B_2), pyridoxine (B_6), pantothenic acid (B_3), biotin, cyanocobalamin (B_{12}), choline, folic acid (B_{10} or B_{11}), inositol, niacin (nicotinic acid or nicotinamide) and para-aminobenzoic acid (PABA). Pangamic acid has been purported to be a B-vitamin (B_{15}); however, in contrast to its popular promotion, ample evidence indicates that it is not a *vitamin* and is not needed by nor of any benefit to any animal. All of the B-vitamins are produced by bacteria in the horse's *cecum* and intestinal tract, and are present in natural feedstuffs in more than ample amounts to meet the horse's needs. There is no indication for giving any of the B-vitamins to most normal, healthy horses.

It has been claimed that giving B-vitamins helps decrease anxiety and nervousness in some horses (settles the horse down). However, there is no evidence based on well controlled studies either confirming or disproving this claim. B-vitamins might stimulate appetite and therefore may be beneficial for the horse that is sick or not eating well. With the exception of vitamin B_{12}, there are no body stores of B-vitamins, so that a deficiency could occur if the horse was not consuming enough feed or if the numbers of bacteria in the intestinal tract which produce these *vitamins* were greatly decreased. If the B-vitamins are deficient, the utilization of *nutrients* is impaired. Brewers' yeast is an excellent source of all of the B-vitamins. It will provide ample quantities of them for the horse that is not eating well when 5 lbs/ton (2.5 kg/metric ton) are added to the *grain* mix or 0.5 oz (14 g) are fed to the horse daily.

Performance horses, particularly race horses, are often given additional *vitamins*, especially B-vitamins (most frequently vitamin B_{12})

to increase their performance. However, there is no evidence that this improves performance. It has been shown that the mature horse does not need any vitamin B_{12} in the diet. **Feeding or injecting excess quantities of vitamins or other nutrients has no beneficial effect on performance and may, if anything, be harmful.**[53]

Vitamin C, also called ascorbic acid, is not needed in the horse's diet. More than adequate quantities to meet the animal's needs are produced in the livers of all animals except rainbow trout, Coho salmon, fruit bats, guinea pigs, red vented bulbul birds, and primates (including humans). *Vitamin C* has been claimed to enhance wound healing and to be of benefit in the treatment or prevention of *infectious* diseases, bleeding from the nose upon exertion, and bone problems developing during growth. However, there is no convincing evidence supporting any of these claims, and there is evidence refuting some of them.

FREE-CHOICE CONSUMPTION OF NUTRIENTS

Animals choose feeds containing certain *nutrients* for one of three reasons.

1. A **true appetite,** in which a *nutrient* is eaten according to the animal's needs for that nutrient. The only nutrients for which animals have a true appetite are water, sodium (common salt), and those required to meet the animal's energy needs.

2. A **learned appetite,** in which the animal learns that a given food or *nutrient* results in feelings of well-being or sickness.

3. A **taste preference,** with no relation to nutritional need or learned appetite, but instead due to feed characteristics affecting *palatability.* With the exception of sodium, all other *minerals* and *vitamins,* if available, are consumed according to taste preference. Many diseases, nutritional and non-nutritional, cause a change in taste preference. For example, animals deficient in phosphorus may chew on bones. This has frequently been cited as evidence that they have a true appetite for phosphorus. However, they may also eat dirt and chew on wood and rocks that are poor sources of phosphorus. It has been well proven that animals do not consume vitamins or minerals (with the exception of sodium) according to their needs, and that when they are allowed to consume these *supplements free-choice,* there is much individual variation.[21,32,62] The amount consumed is completely unrelated to dietary needs. Average consumption by a group of animals may, in some cases, by chance only, result in proper levels of mineral intake. However, intake by the individual animal will vary from too little to excessive. **Free-choice, or cafeteria-style feeding of vitamins and minerals, should never be relied upon to meet the animal's**

requirements. The only way to ensure that each animal receives its needed vitamins and minerals is to have the proper amount present in the animal's water or energy-containing feeds. Occasionally, vitamins, minerals, and drugs, are added to the salt. Although animals have a true appetite for sodium, salt intake is erratic and depends upon the sodium content of other feedstuffs, so the consumption of other minerals or vitamins, when added to the salt, may be quite variable.

Chapter 2
Feedstuffs

Two major types of feedstuffs are available for the horse, *roughages** and *concentrates*. Roughages include loose hay, hay pressed into *cubes* or wafers, pasture, *haylage*, and *ensilage*. Concentrates are generally considered to be everything in the *grain* mix; in addition to cereal grains, this may include such things as *protein supplements, vitamins, minerals*, bran, molasses, and *oil (fat)*. Concentrates containing molasses are frequently referred to as "*sweet feeds*." Commercially prepared horse feeds may be composed entirely of concentrates or contain varying amounts of roughage.

DETERMINING NUTRIENT CONTENT OF FEEDS

The *nutrient* content of the most common feeds used for the horse is given in Appendix Table 1, and that for some commercially prepared horse feeds, in Appendix Table 6. The nutrient content of *concentrates* varies little from the values given. In contrast, the nutrient content of *forages* may be quite different in different cuttings, in hay from different fields, and in hay harvested in different years from the same field. Therefore, the values given in Appendix Table 1 for forages should be considered approximations only.

The most accurate way to determine the *nutrient* content of a feed is to have it analyzed. Local extension services can usually provide the information on where to have feeds analyzed and the cost of the service. For an analysis to be of any benefit, the sample analyzed must be representative of the feed being considered, so the most important single step in determining the nutrient content of a feed is proper sampling. The best method of sampling bales or stacks of hay is to use a special hay probe or coring tool that can be used to bore into the hay and remove a sample (Fig. 2–1). The feed analysis laboratory, extension agent, or feed store may have a hay sampling probe available. If one is not available, simple grab samples may be taken from the inside of the bale or stack. Small samples should be taken from several bales

---

*Words in italics are defined in the glossary.

Fig. 2–1. Hay probe used for taking a sample for nutrient analysis. At the top of the figure is the probe, which is sharp at one end to cut into the hay, and has handles at the other end to twist the probe into the hay. Below it is a solid rod used to push the core of hay out of the probe. In the center is a plastic bag containing the hay sample and at left is a cap that is inserted over the sharpened end of the probe when it is not in use.

or stacks and combined for analysis. The sample should immediately be put into an air-tight plastic bag to prevent moisture evaporation (Fig. 2–2).

Feeds should be analyzed for moisture, crude *fiber*, crude *protein*, calcium, and phosphorus content. Moisture and crude fiber content indicates the quality of the feed, and protein, calcium, and phosphorus content is needed to formulate the *ration* or to determine whether the feed meets the horse's requirements for these *nutrients*.

Some laboratories report the *nutrient* content of feeds as the amount present in the feed *dry matter*, while others report it as the amount present in the feed as fed. If it isn't stated, it will generally be the amount present in the feed as fed, or as the laboratory received it. If the laboratory results are reported as the amount of nutrients present in the feed dry matter, these values must be converted to the amount present in *air dried* feed to compare with the horse's requirements as given in Appendix Table 2, or to compare with the amounts given for that type of feedstuff in Appendix Table 1. This may be done as shown in the following examples.

Fig. 2–2(A,B). Taking a hay sample for nutrient analysis using a hay probe. After the probe is bored into the hay (A), the core of hay it contains is removed. Several cores of hay should be taken, put into a plastic bag (B), sealed, and sent to a feed analysis laboratory. Local agricultural extension agents or veterinarians usually know the location of the nearest laboratory. The minimum analysis required to formulate a feeding program for the horse is for crude protein, calcium, and phosphorus. Analysis for moisture, and either crude fiber, TDN, or energy, may also be necessary to assess the quality of the hay.

Example 1

A laboratory analysis reports that a sample of alfalfa hay contains 15% crude *protein*, 35% crude *fiber*, 1.20% calcium, and 0.25% phosphorus on a *dry matter basis*, i.e., in the feed dry matter or on a moisture-free basis. The amount of these *nutrients* in the *air dried* feed would be the values reported multiplied by 90%, since air dried feed contains 90% dry matter and 10% moisture or water. Thus, this alfalfa hay would contain 13.5% crude protein (15% × 0.90), 31.5% crude fiber (35% × 0.90), 1.08% calcium (1.2 × 0.90), and 0.225% phosphorus (0.25% × 0.90) on an air dried basis. These values may then be compared with those given in Appendix Table 2 to determine whether they meet the horse's requirements for these *nutrients*. These values may also be used to compare with the amounts given for that type of feedstuff in Appendix Table 1. In this example, this hay would be similar to full bloom alfalfa hay and therefore is a fairly poor quality hay.

Example 2

A laboratory analysis reports that a sample of green grass contains 60% moisture and 5% *protein* as received, or as fed. The grass, therefore, contains 40% *dry matter* (100% − 60% moisture) and 5% ÷ 0.40 or 12.5% protein in its dry matter. If the grass were cut as hay and *air dried*, its moisture content would decrease to about 10% and it would therefore contain 90% dry matter. The amount of protein present in the air dried grass would be 12.5% × 0.90 or 11.25% protein on an *air dry basis*. This amount (11.25%) may then be compared with the amount of protein needed as given in Appendix Table 2.

ROUGHAGES

Roughages are fed to the horse in three forms, hay, silage or haylage, and pasture. Hay is prepared by cutting the forage, raking it into windrows, and allowing it to dry in the field. If it doesn't dry to less than 13% moisture content, it may mold during storage. Hay may be piled into stacks, baled, or pressed into cubes, wafers, or pellets for storage. Most hay available commercially is baled, cubed, or pelleted (Fig. 2–3).

Questions often arise concerning which cutting of hay is best. The first cutting will be high in nutritional value if harvested at the proper time and if it does not contain large numbers of weeds that have grown up over the winter and spring. However, first-cutting hay frequently contains more weeds, and in many areas, putting it up at the proper time without it getting rained on after cutting is more difficult than it is for later cuttings of hay. The cuttings from growth that occurs during the hottest part of the growing season are generally the fastest growing. Fast growth results in more stem and less leaves, which decreases the nutritional value. Later cuttings generally have the highest leaf and *nutrient* content, the least weeds, and the best opportunity of being put up without being rained on after cutting and, therefore, the highest feeding value. **However,** in selecting hay, the major consideration is not the kind of hay or the cutting, but the quality of the hay and the nutritional value in relation to its cost (see Determination of Least-Cost Feedstuff in this chapter).

Types of Roughages

The three major types of roughages are *legumes*, grasses, and *cereal grain* hay (Figs. 2–4, 2–5, 2–6). The major legumes are alfalfa,

Fig. 2–3(A,B). Protect hay from the weather during storage. A 3-sided hay shed, as shown here, is adequate and relatively inexpensive. It will pay for itself within a few years in preventing weather-induced losses in the feeding value of the hay. If a hay shed or inside storage facilities are not available, waterproof canvas spread over the top and partially down the sides of the stack, and peaked so water will run off may be used. Plastic, instead of a waterproof canvas, is not recommended. Plastic will frequently puncture and allow water into the hay, but prevent its evaporation, resulting in more spoilage than would have occurred if the hay had been left uncovered.

Fig. 2–4. Alfalfa hay. Note the leaves and stalks which distinguish it from grass hay.

Fig. 2–5. Grass hay. Note the fine stems, long leaves, and heads (below and to the right of the thumb) which distinguish it from legumes.

Fig. 2–6. Oat hay. Note the heads which contain the grain.

(lucerne), clover, birdsfoot trefoil, and lespedeza. A variety of grasses are commonly fed and include timothy, brome, orchard grass, Bermuda grass, bluegrass, fescue, wheatgrass, reed canary grass, blue grama, and many others. Cereal grains, in which the *grain* has not been harvested, may also be cut and used for hay. The more grain it contains, the greater its nutritional value. A good quality cereal grain hay is similar nutritionally to a good quality grass hay; however, if the heads or grain are lost, only *straw* remains, which is poor feed but a good bedding (see discussion of bedding in Chapter 10). Green legume pastures may cause bloat in cattle or sheep, but not in horses. A pasture containing both grass and legumes provides the greatest quantity and balance of total *nutrients* and the longest grazing season.

Alfalfa and other *legumes* are higher in nutritional value than equal quality grass *roughages*, generally containing twice as much *protein* and three times as much calcium, and being much higher in all of the *vitamins*. Because of the greater amount of these *nutrients* in alfalfa, and the increased needs for these nutrients during growth, lactation, and the last three months of pregnancy (Appendix Table 2), alfalfa is preferred during these periods, if it is good quality and available at a reasonable cost as compared to equal quality grass hay. If good alfalfa is not available, grass or cereal grain hay may be fed, and the

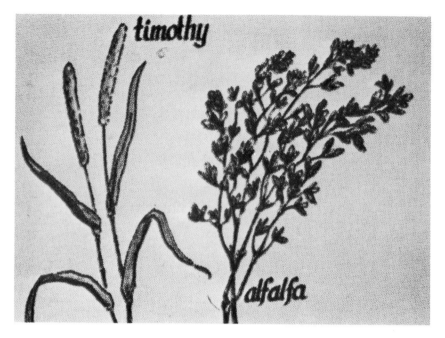

Fig. 2–7. Illustration depicting the differences in appearance between a grass, such as timothy, and a legume, such as alfalfa. In the grass, note the long leaves, their attachment, fine stems, and heads. The stalks of the legume are much coarser, and the leaves less firmly attached. (Courtesy of Dr. R. Seaney, Cornell University.)

additional nutrients needed may be added to a grain mix as described in Chapter 3.

The leaves are less firmly attached to the stem in alfalfa than in grass. This can result in a greater loss of leaves from alfalfa if it is not cut at the proper stage of maturity or handled properly after cutting. Since leaves contain two thirds of the *energy,* three fourths of the *protein,* and most of the other *nutrients* present in *roughages,* the loss of leaves greatly decreases nutritional value. The firmer attachment of the leaves and the difference in their size and shape in grass hay makes leaf loss less of a problem (Fig. 2–7). This permits more latitude in the cutting and processing of grass hays without greatly decreasing their nutritional value. In addition, grass hay is usually less dusty, which makes it cleaner to feed and results in less coughing and *pulmonary emphysema* or *heaves.*

HAY QUALITY

Characteristics of good hay are that it is (1) free of mold, dust, and weeds, (2) leafy with fine stems, (3) soft and pliable to the touch, (4) harvested at an early stage of maturity, and (5) a bright green color rather than a yellow or brown (least important, although worth consideration). Early cutting, rapid drying, and minimum handling will decrease the loss of leaves. As the plant matures, its digestibility and *energy* and *protein* content decreases. Just prior to the time *legumes* flower (the bud or vegetative stage), and when grasses begin to show heads through the sheath (boot stage), their leaf development has been completed and they should be cut for hay (Fig. 2–8). At this time, fields of grasses will begin to change from a deep green to a slight gray as the heads begin to appear. Legumes consist of about one-half leaves and one-half stems at the bud stage. Allowing the plant to stand after first flowering or past the boot stage increases crude *fiber* and reduces crude protein ½% per day, and decreases digestible energy about ¾% per day. As legumes mature from early bloom to full bloom, and grasses mature from the boot stage to complete heading out, one half of the protein and one third of the energy is lost. This is also true of pasture *forage*. As forage quality decreases, so does *palatability* and digestibility. These factors result in decreased total digestible *nutrient* intake and feeding value. Grass hay that has heads

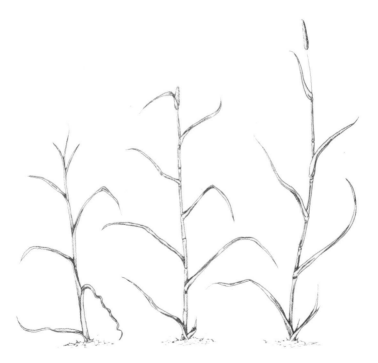

Fig. 2–8. Growth stages of a typical grass, timothy. The grass should be harvested no later than the boot stage, which is when the head begins to show through the sheath (center figure). (Courtesy EQUUS, 32, (6), 1980.)

Fig. 2–9. Moldy hay. A white mold was present not only on the edges of this bale but throughout, and the hay was dusty. Moldy hay is unpalatable, contains fungal spores that cause heaves, coughing, and bleeders, and may contain mycotoxins. White dust in hay is usually fungal spores.

over ½ inch (1 cm) long and a legume hay that is past full bloom are too mature to make a good feed.

The most accurate and reliable way to determine hay quality is to have it analyzed, at least for moisture, *protein*, and *fiber*, as described earlier in this chapter. If the hay contains less protein or more fiber than the amounts given for that type of feedstuff in Appendix Table 1, it is a below average quality feed. It is important that this comparison be made on an equal moisture content basis, as described previously in this chapter. A crude fiber content greater than 33% indicates a poor-quality hay. If the moisture content of the feed is greater than 13%, it may become moldy during storage and should not be fed. Moldy feed may cause *chronic* coughing, *heaves*, and *bleeders*, and may contain *mycotoxins*, which can cause abortion and death. In addition, moldy feeds may be unpalatable (Fig. 2–9).

If the only feed available is of poor quality, feed lots of it. This allows the horse to sort through it and eat only the best portions. If smaller amounts are fed, the horse is forced to eat the poorer quality portions of the feed.

HAY CUBES

Hay is usually fed in loose form, but it may be pressed into wafers, *cubes,* or pellets (Fig. 2–10). These have a number of advantages and

Fig. 2–10. Hay pressed into cubes. Cubes are 1 to 1¼ inches (3 cm) square and 1 to 3 inches (3 to 7 cm) long.

disadvantages compared to loose or long-stem hay. The advantages are as follows:

1. Less wastage by the horse. With cattle, loss of baled hay during storage and feeding in racks is 9%, as compared with 4% loss when pelleted or cubed hay is fed. When loose hay is fed on the ground, losses may be 18 to 30%. Losses similar to these might be expected for horses.

2. Less storage space is required. A ton of hay *cubes* occupies 60 to 70 cubic feet, as compared with 200 to 330 cubic feet for each ton of baled hay and 450 to 600 cubic feet for each ton of loose hay.

3. Transportation costs may be reduced.

4. If made properly, there is less dust when hay cubes or wafers are eaten. This is an important advantage for horses, in which the dust or *fungal spores* in loose hay may cause *heaves*, coughing, and *bleeders*, (see Chapter 4).

5. The loss of leaves from alfalfa by handling or by the horse is less with alfalfa *cubes* or wafers than with loose alfalfa. Thus, when equal quality alfalfa is used, greater nutritional value is received from cubes or wafers.

The disadvantages of hay *cubes* and wafers are as follows:

1. If they are too soft and crumbly, they will break into fine material that is dusty, easily lost, or may cause digestive problems. If they are

well made, this is not a problem unless they get wet. For this reason, they must be protected from the weather. If *roughage* is cubed or wafered when it is too dry, it will be soft and crumbly, whereas if it is too wet, mold growth and spoilage occur.

2. They may be more difficult than loose hay to feed by hand, although they are generally easier to feed by machine.

3. *Wood chewing* may be increased when *cubes* or wafers are fed as the only source of *roughage* (see Chapter 4).

4. It is difficult to determine their quality without a laboratory analysis. If there is any question about their quality, they should be analyzed, at least for moisture, *protein*, and *fiber*. The results of the laboratory analysis should be compared with the amounts given for that type of feedstuff in Appendix Table 1. If they contain less protein or more fiber than the amounts given, it means that they are below average quality feeds. Be sure that this comparison is made on an equal moisture content basis, as described earlier in this chapter.

SILAGE AND HAYLAGE

Silage or *haylage* are two additional forms of *roughage* that may be fed. If good quality, they are a highly nutritious, palatable *forage* for the horse. They should be free of mold. Spoiled or moldy silage, haylage, or any feed can be poisonous to horses. Since these are fermented feeds, some horses may be reluctant to eat them until they become accustomed to them. They should not constitute over one half of the roughage in the horse's *ration*. Since silage and haylage contain about one-third *dry matter* and two-thirds water, and hays contain 90% dry matter, it takes about 3 lbs of silage or haylage to replace 1 lb of hay.

PASTURE

Pasture management and seeding differ from area to area. Therefore, it is recommended that the local extension service be contacted for the best procedures for that area. The following are given as examples of the factors that should be considered. Some of these factors may not be correct or optimum for your specific area.

The horse will eat, trample, or damage *forage* that is the equivalent of at least 1000 lbs (454 kg) of hay per month. Forage production of pasture in most areas occurs during a 5- to 6-month period. During this period, one acre of good, improved, irrigated pasture may yield the equivalent of 5 tons (4.5 MT) of hay. An acre of good, improved coastal Bermuda grass may yield 5 to 7 tons (4.5 to 6 MT) of hay without irrigation in some areas. Thus, one acre (0.4 ha) of these types of pasture would support two horses during this period. In contrast, 30 to 60 acres (12 to 24 ha) of dry range pasture may be needed to support a single horse for one year.

It is best not to allow horses on the pasture during and shortly following irrigation. Adequate time should be allowed for drying of pastures to minimize trampling, plant injury, and soil compaction. Pastures should be irrigated at intervals that do not permit plants to be stressed. The soil should not become dried out below the top two inches. The approximate amount of total annual water needed (rainfall, snow, and irrigation water) is 24 to 36 inches (60 to 90 cm).

Forage production of some types of established pastures can usually be greatly increased with nitrogen fertilization at the beginning of the growing season and midsummer.[58] *Legumes* will also provide additional nitrogen for grass growing with them and therefore increase grass growth. In addition, legumes improve the physical condition of shallow clay soils, resulting in further improvement of grass growth on this type of soil. Legumes will grow during periods in which warm season perennial pasture grasses grow little or not at all and therefore they prolong seasonal forage production from the pasture. For these reasons, seeding legumes with grasses or into established grass pastures is generally beneficial.

Before seeding a pasture, a soil sample should be taken. If the soil is deficient in phosphorus, an adequate supply of phosphorus (based on the soil test) should be applied and plowed under prior to seeding. In addition, 0 to 50 pounds of phosphorus (P_2O_5) per acre (0 to 50 kg/ha) and nitrogen (the amounts of both are based on the soil test) should be applied just below the seed during planting. A soil test is also valuable in determining the species of grasses best suited to the area.

Grasses differ as to productive capacities. The most productive on irrigated pastures is usually orchard grass, followed by tall fescue, bromegrass, and intermediate or pubescent wheatgrass. Differences in productive capacity, however, exist in different areas and with different soil types.

The pasture grasses that generally produce the most forage and provide the best nourishment to the horse in different geographic regions of the United States are as follows:

Florida, South Georgia, Gulf Coast—pangola, bahia, coastal Bermuda

Middle Atlantic—coastal Bermuda (southern area), orchard grass, bluegrass, white and red clover

Northeast—redtop, orchard grass, reed canary grass

Midwest—smooth brome, buffalo grass, bluestem, grama

Southwest—coastal Bermuda, ryegrass

Northwest—fescue, bent grass, bluestem, grama, crested wheatgrass

Far West—orchard grass, coastal Bermuda, lovegrass, Rhodes grass, rescue grass

Generally only one, or at the most two species of grass should be planted, since grazing will usually eliminate all but one. Alfalfa, clover, or other *legumes* should be seeded with the grass. Seeding rates should be based on the amount of pure-live seed. The fraction of pure-live seed can be determined by multiplying "purity" by "germination." The amount of bulk seed to use can then be determined by dividing the recommended amount of pure-live seed by this fraction (Table 2–1).

On pastures of bluestem, smooth brome, crested wheatgrass, or *legumes* such as alfalfa or clover, or on native pastures in the Rocky Mountain–Great Plains area of the United States, only 50 to 60% of the *forage* should be removed by grazing. Grazing in excess of this is harmful to the plants and slows regrowth, so that over several years the average yearly yield is lowered. In contrast, Kentucky bluegrass and Bermuda grass, which are the most important pasture grasses in many southern states and the southwest, do best when grazed intensively and cropped closely. Once the pasture is eaten down, the animals should be removed and the pasture irrigated quickly. Kentucky bluegrass and Bermuda grass should be grazed at early flowering or before, at which time their feeding value is the greatest and they may contain as high as 20% *protein*. Leaving a high percentage of these forages unfed will result in large areas of the pasture grass becoming over-mature, unpalatable, high in *fiber*, and parasite-infested. The animals then tend to graze only in the areas that have been closely cropped, and the useful size of the pasture is reduced. To

TABLE 2–1
PURE-LIVE SEED RECOMMENDED FOR DRILL SEEDING PASTURES[58]*

Pasture			
Irrigated		Dryland	
Type of Seed	lbs/acre or kg/hectare	Type of Seed	lbs/acre or kg/hectare
Orchard grass	3	Pubescent wheatgrass	10
Tall fescue	8	Crested wheatgrass	5
Bromegrass	14	Western wheatgrass	8
Alfalfa or clover	8		

*Broadcast seeding requires 50% more seed than drill seeding and results in an inferior stand. See your local extension service for adjustments in the amounts given, and types of forages recommended for your specific area.

correct this once it occurs, cut the over-mature forage close to the ground and scatter manure with a wire drag or harrow.

On small, dryland acreages, a shelter and exercise corral is useful. The remainder of the area can be seeded and only limited grazing permitted so that it is never grazed to less than 2 inches in height (5 cm).

If there is adequate acreage and two to four pastures are available, continuous or rotational grazing may be used. Rotational grazing provides the best opportunity to obtain maximum yield (Fig. 2–11). The horses should be allowed on the pasture shortly before plant growth is complete. For alfalfa, this would be in the prebud stage and at about 8 inches (20 cm) tall. When the *forage* is eaten down to 3 inches (7.5 cm), the horses should be removed. In the fall, as the plant's dormant season nears, it is preferable to leave 4 inches (10 cm).

If *legumes*, such as alfalfa or clover, are being seeded with a grass (which is recommended), the total amount of legumes should not exceed 25% of the seed mix. For example, if orchard grass and alfalfa were used, the seed mix should contain 0.75 × 3 or 2.25 lbs (kg) of orchard grass, and 0.25 × 8 or 2 lbs (kg) alfalfa (the pounds of seed needed, 3 and 8, are given in Table 2–1).

Tall fescue has good tolerance to wet or alkali soils (see Chapter 4 on fescue toxicity). Orchard grass produces excellent quality *forage*, but will not tolerate drought, wet, or alkali soils. If full-season

Fig. 2–11. Dividing an acreage into several pastures so that rotational grazing of each pasture may be used provides the best opportunity to obtain maximum forage production. (Courtesy of Moondrift Farm, Fort Collins, CO.)

irrigation water is not available, smooth bromegrass, intermediate wheatgrass or pubescent wheatgrass may be best.

Irrigated pastures may be seeded from spring through fall; however, in areas that have cold winters the ideal time is spring. Late fall seedings generally do not become sufficiently established to survive a cold, dry winter. Seeding depths should be ¾ to 1 inch (2 to 2.5 cm) in sandy soils and ¼ to ½ inch (0.6 to 1.2 cm) for small seeds in ideally textured soils.

New plantings should be irrigated frequently to keep the seed moist. After emergence, the seedlings should be well watered. It is important to control weeds by spraying or mowing when pastures are reseeded or becoming established. Doing this before perennial weeds bud and before annual weeds seed will give much better control and keep the pastures from becoming reinfested longer. After reseeding, the pasture should not be grazed during the first season after planting and not until it is 6 inches (15 cm) tall during the second season.

For seeding dryland pasture, a sorghum *grain* (milo) stubble, prepared the year prior to seeding, makes a good seedbed and is recommended. Nurse crops or small cereal grain stubble are not as good. Sorghums, Sudan grass, and sorghum–Sudan grass hybrids, however, are not recommended for grazing by horses. Horses grazing these grasses, or eating them as *silage* or *haylage,* may develop a cystitis syndrome or inflammation of the urinary tract that may be fatal. Hays from these grasses, properly cured and stored, can be safely fed to horses, but are generally a poor quality *roughage.* Late winter through early spring is the best time for seeding in most areas. Grazing should not be allowed during the year of seeding. During the next two or three years, it is best not to graze it for one year. Weeds should be mowed or sprayed frequently to prevent shading of the grass and competing for moisture. Any one of three wheatgrasses, pubescent, crested, or western, are recommended for seeding dryland pasture.

Overseeding established perennial warm-season grass pastures with cool-season annuals or *legumes* can provide high quality *forage* during the winter months and greatly extend grazing time. Legumes will also increase the amount of growth of grass growing with it. Successful overseeding of established pastures can usually be obtained by the following:

1. Removing top growth or warm-season grass by heavy grazing or mowing just prior to seeding or cool weather in the fall.

2. Disking the area sufficiently to loosen top soil in order to get seed coverage. This will also retard growth of the warm-season grasses and allow the new seed to become established. Bermuda grass sods can be disked more heavily than bunch grass types such as Klein grass or buffelgrass.

3. Applying fertilizer, based on the results of a soil test.

4. Seeding *legumes* or cool season annuals, such as ryegrass, or small *grains* such as wheat, oats, rye, or barley. Ryegrass and legumes can be broadcast over the surface. Small grains should be drilled in. In either case, some type of drag or harrow should be dragged across the area after seeding so that good seed coverage is obtained.

5. Withhold grazing until the newly seeded *forage* is well established.

CONCENTRATES

Concentrates are feeds that provide a high concentration of dietary energy and, therefore, are low in crude fiber content (under 18%). The major ones fed to the horse are cereal grains, protein supplements and molasses.

As a general rule, *concentrates* should never make up over one half of the total weight of feed eaten by the horse. It is possible to feed *rations* containing as high as 80 to 90% concentrates if the amount of concentrates in the ration is gradually increased over several weeks. However, there is an inherent danger in feeding excessively high amounts of concentrates. If feed consumption is reduced for any reason (such as the horse being sick or affected by a weather change), the horse may *founder* when it later eats the amount of concentrates that it was previously accustomed to. In addition, if the accustomed daily exercise of a horse eating these quantities of concentrates is reduced, *azoturia* may occur the next time it is used, particularly if the horse isn't warmed up slowly. It has been recommended that young horses may be fed rations containing 67% concentrates.[50] However, excess concentrate intake is a major cause of *epiphysitis* and bone problems in the growing horse (see Chapter 9), and is the most common error made in feeding the growing horse. For these reasons, the amount of concentrate in the ration should never make up over one half of the total weight of feed eaten by any horse. Thus, the total ration for the horse may consist of 0 to 50% concentrates and 50 to 100% roughages.

Cereal Grains

Any of several *cereal grains* may be fed to horses. The *nutrient* contents of the most common grains available are given in Appendix Table 1. In contrast to *forages*, the nutrient contents of the cereal grains vary little from the values given. These values, therefore, may be used in formulating *rations* for the horse. A laboratory analysis usually is not necessary. Cereal grains are fed primarily as a source of *energy*. Therefore, cost in relation to *energy* provided should be the major factor in selecting a cereal grain. Methods for comparing the value of feedstuffs are discussed later in this chapter. Most horses prefer the feed that they are accustomed to; therefore, grain intake is often decreased when the type of grain fed is changed.

Oats is the most popular *cereal grain* fed to horses (Fig. 2–13). On an *energy* basis, oats are generally the most expensive cereal grain. In addition, the quality of oats often varies more than other cereal grains, because of variability in the amount of *hulls*. Although dehulled oats, "race horse oats," "jockey oats," or heavy oats (oat *groats*) can be used, they are expensive with respect to their nutritional value. Because oats are lower in energy than other cereal grains, more must be eaten to produce *founder* or digestive problems. However, in contrast to what is occasionally stated, horses will founder on oats if a sufficient quantity is consumed.

Corn or *maize* is commonly fed, and is a good feed for horses (Fig. 2–12). It has been demonstrated in controlled studies that there is no difference in *gastrointestinal* disorders (diarrhea, *colic* or *founder*) in horses in training that are fed corn as compared with those fed oats, with either *grain* making up 40% or 60% of the total ration.[95] Corn is higher in both *energy* content and density than oats, and contains twice as much energy as an equal volume of oats. Because of this, some horsemen feel corn has a tendency to cause obesity or make a horse too spirited ("high"). If equal volumes of corn and oats are fed

Fig. 2–12.   Corn or maize—cracked, flaked or crimped, and whole.

this is true, since the horse is receiving twice as much energy from the corn. However, if equal amounts of energy, not equal volumes, are fed, corn does not have any greater tendency to cause obesity or make a horse too spirited than other *cereal grains.* How spirited a horse acts is directly related to how good it feels, and how much energy it has consumed and needs to use. Corn is often the least expensive grain available. This is particularly true when its energy content with respect to price is considered. This is the reason that corn is so frequently the major ingredient in many cattle, sheep, and swine *rations.* Contrary to popular belief, corn is not a *"heating feed."* It is sometimes fed during the winter and not during the summer because of this mistaken belief. The heat produced in the digestion, absorption, and utilization of corn is one third less than that produced from oats. Forty-one percent of the gross energy in corn is given off as heat, as compared with 66% of the gross energy in oats. However, because corn has a high energy density and because energy needs are increased during cold weather, it is a good winter feed (see Chapter 5).

Barley is a good *cereal grain* for the horse and may be fed as the only grain in the *ration.* It is intermediate between corn and oats with respect to *energy* content. **Wheat** and **milo** (sorghum grain) may also be fed to the horse (Fig. 2–13). Wheat, because of its gluten content, should not constitute more than one half of the grain ration. Wheat gluten gives wheat flour its sticky consistency, which makes it so valuable to the baking industry. This sticky consistency is a disadvantage in feeding, but is not a problem unless excessive amounts are fed. **Rye** may also be fed to the horse, but should not make up over one third of the grain ration because of poor *palatability.*

Wheat bran (Fig. 2–14) is commonly added to the *grain ration* or fed as a hot bran mash. The only benefit of feeding hot bran mashes is that it results in the consumption of water. When the environmental and therefore water temperature is quite cold, water consumption greatly decreases. Obviously it decreases even further if the water freezes over and the horse has no other water available. This, unfortunately, is not an uncommon occurrence. Inadequate water consumption may result in feed becoming impacted in the intestinal tract, which results in *colic.* Feeding hot bran mashes increases water consumption and assists in preventing impactions from occurring. This is the source of the belief that hot bran mashes or bran prevents colic. However, installing heated waterers and allowing the horse free access to all the water it wants is more practical and easier for the feeder and better for the horse. There is no evidence to indicate any benefit from hot bran mashes if adequate, cool to lukewarm water is readily available.

In contrast to popular belief, bran does not have a laxative effect, i.e., it does not increase fecal water content, and therefore, soften the

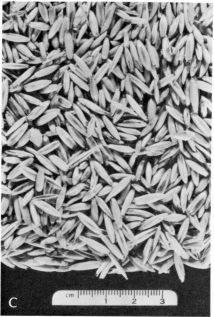

Fig. 2–13(A,B,C). Cereal grains. These are sorghum or milo (A), wheat (B), and oats (C). Oats are similar in appearance to barley.

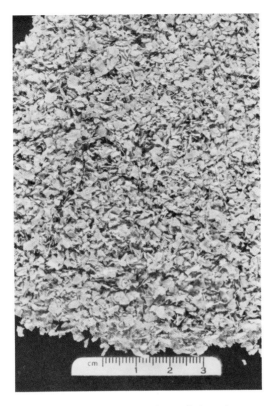

Fig. 2–14. Wheat bran, often called just bran.

stools.[32] It does, however, increase the amount of feces excreted because it is a bulky, poorly digested, low *energy* feed. For the same weight of feed, bran provides about 12% less energy than oats, and 25% less energy than corn or wheat (Appendix Table 1). For the same volume of feed, corn and wheat provide four times more energy and oats two times more energy than bran. Although bran is higher in *protein* and phosphorus content than other *cereal grains*, it is not a practical means of adding these *nutrients* to the *ration*. If additional protein or phosphorus are needed, it is much more practical, and less expensive, to add protein or phosphorus *supplements* rather than bran to the ration. Bran is palatable and there is certainly no harm in feeding it; however, I see little reason to do so. There is no evidence based on controlled studies either supporting or refuting the belief that feeding bran treats or prevents *colic*.

PROCESSING AND STORAGE

Cereal grains may be fed whole or processed. Different processing methods include cracking, rolling, *crimping,* steam flaking, or micronizing (Fig. 2–15). Heating during processing, if not excessive, increases the digestibility and therefore the feeding value of cereal grains for the dog, and it also may for the horse. *Protein* digestion by the horse is 2 to 3% higher from both oats and milo when they are micronized than when they are crimped.[78] Cereal grains should never be finely ground for the horse. Fine grinding increases digestive problems and feed loss, and does not increase feed utilization above that obtained with coarse grinding, cracking, rolling, or crimping. For the horse that bolts its grain, crimping the grain may prevent this. The smaller and harder the cereal grain kernels, the more processing increases its feeding value. Processing increases the digestibility of rye, wheat, and milo by about 15%; corn, 7 to 9%; and oats and barley, 2 to 5%. This means that if rolled oats can be purchased for not over 5% above the cost of whole oats, they are a good buy. If processing costs more than it increases the digestibility, it isn't worth the added cost. The exception would be when the grain is for the older horse with poor teeth.

Fig. 2–15. Oats. From left to right, whole, rolled or crimped, and steam flaked. Either rolling or steam flaking increases the feeding value of oats 2 to 5%, but decreases the length of time they should be stored.

Processing *cereal grains* does not predispose to more rapid gas production in the stomach and as a result to gastric rupture, as some have claimed. Cracking the cereal grain by processing does, however, decrease stability during storage. Cereal grains that do not include broken kernels, that contain less than 13% moisture, and are free of rodents, may be stored for years with little loss in nutritional value. Cereal grains containing broken kernels, such as processed grains, become oxidized during prolonged storage and develop a stale flavor that decreases their *palatability*. Mold growth also occurs more readily in a grain containing broken kernels. The higher the temperature and humidity, the less time required for oxidation and mold growth to occur. Molds or fungi may produce *mycotoxins* that are detrimental to the horse, and may cause death. Moldy feeds should never be fed to the horse. Antioxidants, which are present in some

Fig. 2–16(A,B). Vermin- and moisture-proof grain bins should be used. Metal containers such as large garbage cans or wood bins, with tight lids, are excellent for limited feed storage. These should be kept in a feed room equipped with a latch so that a horse cannot get in. Mice and other vermin rapidly eat through sacks and plastic garbage cans. Hopper bottom bins are excellent for storing larger quantities of feed, and allow the purchase of grain in bulk. Purchasing grain in bulk greatly decreases feed cost as compared with purchasing it in sacks. However, if the grain kernels are broken, as happens in processing, not more than several months' supply should be purchased at one time, to prevent grain from becoming stale and losing palatability and nutritional value prior to feeding.

commercial feeds, may be added to a grain mix to prolong the time before oxidation or mold growth occurs.

For the person with no more than a few horses, galvanized garbage cans with good tight fitting lids work well to store *grain* (Fig. 2–16A). Rodents gain easy access to grain stored in sacks or plastic garbage cans, and grain left in sacks may become moist and moldy. For those with a number of horses, galvanized metal bulk storage containers are best (Fig. 2–16B). Feed costs are greatly reduced when bulk, rather than sacked, grain or grain mixes are purchased. However, moisture condensation in large metal feed bins may result in some feed spoilage. This occurs particularly in feeds containing greater than 12% moisture, or during warm weather in humid climates.

GRAIN QUALITY

Grains, like all feeds, vary in quality. Those fed should be free of mold and foreign material, and contain a moisture content low enough to prevent excessive heating during storage (13%). Government grading standards are one of the best indications of grain quality. The best quality grain is graded U.S. No. 1, followed in order by Nos. 2, 3, 4, and for some types of grain, 5. U.S. Sample grade is the poorest quality. The higher the weight per bushel (Table 2–2) and the lower the moisture content, foreign material, broken and damaged kernels, and discoloration, the better the quality of the grain, and the lower its grading number.

TABLE 2–2
MINIMUM WEIGHT PER BUSHEL IN THE UNITED STATES STANDARDS FOR GRAINS[96] *

Grade	Barley	Corn, Rye and Soybeans	Oats	Milo (Sorghum)	Wheat†	Wheat‡
U.S. No. 1	47	56	36	57	58	60
U.S. No. 2	45	54	33	55	57	58
U.S. No. 3	43	52	30	53	55	56
U.S. No. 4	40	49	27	51	53	54
U.S. No. 5	36	46	—	—	50	51

*Amounts given are lbs/bu; to convert to kg/kl, multiply values given times 12.9, and to convert to lbs/qt, divide value by 32.
†Hard red spring, or white club wheat.
‡All other wheat.

Protein Supplements

As shown in Tables 6–1 and 8–1, many feeds don't contain enough *protein* to meet the requirements of the lactating mare and the growing horse. The additional protein needed is usually provided by adding a feed that is high in protein to the *concentrate* mix. Protein *supplements* may be of animal or of plant origin. Those of plant origin are more commonly fed to livestock.

The major plant-origin *protein supplements* are the oilseed meals. These are by-products of the extraction of *oil* from soybeans, cottonseeds, flaxseeds (linseed meal), peanuts, safflower seeds, sesame seeds, sunflower seeds, rapeseeds, and coconut (copra meal). Soybean meal is the most widely used protein supplement for livestock, followed by cottonseed meal. The oilseed meal protein supplements contain from 32 to 50% protein, and are generally referred to according to their protein content; for example, 44% soybean meal indicates that it contains 44% protein in the form in which it is fed.

Linseed meal occasionally is fed because it is thought that it contains a higher *oil* or *fat* content, which gives the horse a glossy hair coat. Although this was true when the old, or mechanical, process was employed to extract oil from flaxseeds, the new or solvent process is used now almost exclusively. This process removes more of the oil, so that the remaining linseed meal doesn't have any advantage over the other oilseed meals in giving the horse a glossier hair coat. In addition, linseed meal is low in *lysine*, which is needed for optimum growth.

Soybean meal contains the highest quality *protein* and the highest amount of *lysine* of any of the plant-source protein *supplements*. **Soybean meal is the preferred plant-source protein supplement for the growing horse.** (See Chapter 1 for a discussion of protein quality and its importance for the growing horse.) Although palatable, it is not generally as palatable as **cottonseed meal.** However, protein quality is lower in cottonseed meal. In the past, some cottonseed meal contained a toxic substance called gossypol. Gossypol is toxic to young horses, and especially to pigs in which it may cause blindness and abortion. Gossypol is rarely a problem now, however, because of improved cottonseed plants and processing methods.

Peanut meal is a palatable *protein supplement,* but tends to become rancid when stored for too long, especially in warm, moist climates. It should not be stored longer than six weeks in the summer and two to three months in the winter. It is even lower in *lysine* than linseed meal, and therefore should not be used as a protein supplement for the growing horse. **Rapeseed meal** is a high-quality protein supplement comparable to soybean meal. However, it is rather unpalatable. In addition, it contains several compounds that interfere with thyroid function. For this reason, it is advisable not to feed it to brood mares, and to limit intake for other mature horses to 2 lbs (1 kg) per day. **Safflower meal** generally contains about 15% *fiber*, twice that of other oilseed meals, and therefore is lower in utilizable *energy*. Also, its palatability is low. **Coconut,** or **copra, meal** is lower in both protein content and protein quality than the oilseed meals. Decorticated **sunflower meal,** although lower in lysine content and *palatability* than soybean meal, is equal in nutritional value to soybean meal for the mature horse.

Additional plant-protein *supplements* are obtained as by-products from *grain* milling, starch production, brewing, and distilling, e.g., brewer's grain. **Brewer's grain** may be referred to by the name of the brewer, as in Coors pellets. Most of these industries use the starch present in the grain, then dispose of the residue, which contains a large portion of the *protein* of the original plant seed. Brewer's grains generally contain 25 to 30% protein and are palatable, but like all protein derived from cereal grains, are some of the poorest quality available. Supplements such as corn gluten meal or corn gluten feed are low in the essential amino acids, lysine, tryptophane and methionine, which are needed for growth. This is also true for protein supplements derived from other cereal grains. Therefore, they should not be used as a source of protein for the growing horse, but may be used for the mature horse.

Animal-source *protein supplements* are generally derived from inedible (i.e., not for human consumption) tissues from meat-packing or rendering plants, and from surplus milk or milk products. Protein supplements from fish, poultry, and eggs, and by-products of their processing may also be used. Although animal-origin protein supplements may be excellent feeds, there are several problems associated with them. Most contain large amounts of *fat,* which may become rancid (see Chapter 1). They are excellent media for bacterial growth, so many of them must be processed and stored in a manner to prevent bacterial growth. In addition, they are generally more costly, and many are less palatable to horses than plant-protein supplements. **Dried milk** and milk products, although generally costly, do not share many of the disadvantages of the other animal-source protein supplements. Milk products are an excellent protein supplement for the growing horse because of their high *lysine* content (Table 1–4).

Molasses

Molasses is extremely palatable, and is a good source of *energy*. On a *dry matter basis*, it provides about the same, or slightly less (10%), energy than oats. It is available in both liquid and *dehydrated* forms. Dehydrated molasses contains less than 10% moisture. It is less commonly used than the syrupy, liquid molasses, which contains 22 to 32% moisture. Liquid molasses is added to *concentrate* mixes (1) as an appetizer, (2) to reduce dustiness, and (3) as a binder for pelleting, or to keep *minerals, protein supplements,* and other ingredients from sifting out of a loose concentrate mix. *Fats* or *oils* may also be added to concentrate mixes for these same reasons. Concentrate mixes containing molasses (and rarely, brown sugar, or honey) are often referred to as *"sweet feeds"* (Fig. 2–17). More than 10 to 15% liquid molasses in the mix causes it to cake or become messy. In hot, humid areas, molasses should be limited to 5% of a loose concentrate mix; otherwise, mold may develop. When used as a binder for pelleting, 7.5 to 10% is generally optimum. Too much molasses makes the pellet too soft and chewy, whereas too little makes the pellet crumbly.

There are several different types of molasses: citrus; wood; starch, or corn; and cane and beet molasses, which are by-products of the manufacture of sugar. Cane, or blackstrap, molasses is the most extensively used, followed by beet molasses. Although similar in *energy* content, they differ in *protein* content. Cane molasses contains 3 to 5% protein; beet molasses, 6 to 10%; citrus molasses, around 14%; wood molasses, about 1%; and corn molasses, less than 0.5% protein.

Fig. 2–17. A typical "sweet feed," or grain mix that contains molasses.

COMMERCIALLY PREPARED FEEDS

The three types of commercially prepared horse feeds are (1) *roughage* pellets or *cubes,* (2) *grain* mixes, and (3) pellets containing primarily roughages, with varying amounts of cereal grains, usually 10 to 25%. Grain mixes may be loose, or compressed into pellets or cubes, which are called cake. The loose mixtures generally contain molasses, and are referred to as *sweet feeds.* When feed is pelleted, the size of the pellet isn't important. Pellets about ½ by 1 inch (1 by 2 cm) are generally preferred, although if horses are fed on the ground, larger pellets are better. The pellets should be well-formed and not crumbly. Pelleted feeds are made by grinding the feeds, mixing them with a binder such as molasses to hold them together, and forcing the mixture through a sieve.

The three different types of pelleted feeds may be differentiated from each other by their crude *fiber* content, which is given on the feed tag. The lower the crude fiber content of the feed, the greater its *energy* content and the lower its *roughage* content. Pellets consisting entirely of roughage contain 27% or more crude fiber. Those containing primarily roughage with some cereal grain, contain 18 to 26% crude fiber. Cereal grains contain less than 15% crude fiber.

Dehydrated alfalfa pellets made from fresh-cut alfalfa and containing no *cereal grains* are frequently called *"dehy"* (Fig. 2–18). When good quality alfalfa is used, they are referred to as 17% dehy, which indicates that they contain 17% or more crude *protein.* When poorer quality alfalfa is used, the pellets are called 15% dehy. The 15% dehy contains 33% crude *fiber,* whereas 17% dehy contains 27% crude fiber. The heat used in dehydrating the alfalfa denatures some of the protein, so that as much as one third of it is undigestible. Although the *palatability* of dehy may be poor, it is well tolerated when added to a grain mix. Dehydrated alfalfa pellets should not be confused with sun-cured alfalfa hay that is pressed into *cubes* or pellets. These are not dehydrated, and were described previously with other *roughages.* Nondehydrated, or sun-cured alfalfa cubes or pellets, do not have the disadvantages of dehydrated pellets.

Pelleted feeds consisting primarily or totally of *roughage* may be fed without any additional roughage. However, when this is done, *wood chewing* may greatly increase. This may happen also when hay *cubes* or wafers are fed as the only source of roughage for the horse. If wood chewing occurs when pellets, cubes, or wafers are fed, at least one-half pound of long stem hay/100 body weight/day (½ kg/100 kg) should also be fed.

Pelleted feeds have several advantages and disadvantages compared with nonpelleted *rations.* The advantages are the same as those described for hay wafers or *cubes.* Additional advantages are that (1)

Fig. 2-18. Alfalfa or "dehy" pellets. These are dark green, and may be made from either sun-cured or dehydrated alfalfa.

intestinal fill is reduced, so that as much as 20 to 30% more pelleted feed than loose *forage* can be consumed, (2) all *vitamins, minerals,* and *protein supplements* can be added and will not sift out, allowing for precise control of the ration consumed, (3) the amount of manure may decrease, and (4) pellets are easy to feed to the horse. A decrease in intestinal fill may make some horses look better because they do not have *hay bellies.* The increased amount that can be consumed, the decreased loss of leaves in processing and consumption, and the grinding of the *roughage* in pelleted feeds are important advantages for the *hard keeping* horse, particularly the older horse with poor teeth. Often thin horses of this type that have been receiving loose roughage will put on weight when switched to a pelleted feed.

Feeding pellets may also decrease the incidence of *colic* that can occur in the older horse with poor teeth, and consequently improper mastication of roughage. A complete pelleted feed containing 10 to 25% *grain*, as indicated by an 18 to 26% crude *fiber* content, is recommended for the older horse that has trouble maintaining optimum body weight when fed long stem hay and grain. If additional *energy* intake is necessary to maintain body weight and condition, one should feed cereal grains that have been processed.

The major disadvantage of pelleted feeds, in addition to those described earlier in this chapter for hay wafers or *cubes*, is that they may be more costly than long stem hay and grain rations. However, they may not be if their advantages, such as less waste and less storage space required, are taken into consideration. "Choke" in horses that are greedy eaters is occasionally cited as a disadvantage of pelleted feeds. This is an uncommon problem, but if it occurs, it can be prevented by putting round rocks in the feedbunk so the horse has to eat around them. The rocks should be large enough so the horse can't get them in its mouth, and hard enough so they won't break.

DETERMINING LEAST COST FEEDSTUFF

Feeds should always be purchased by weight, not by volume, since the density of feeds, particularly hay, may be variable. Similar-sized square bales of hay may weigh from 35 to 120 lbs (16 to 55 kg). This emphasizes the importance of buying hay by weight, not by the bale. Even *cereal grains* may vary by as much as 10 lbs per bu (129 kg/kl). The best price for hay can generally be obtained during the growing season. Hay prices routinely increase, often by very large amounts, during the winter, or spring before the first cutting is available. In addition to these considerations, determining the least cost feedstuff can often greatly lessen the cost of feeding the horse.

To determine the least cost feedstuff, the cost of the major ingredient of concern in each feed should be compared. The major ingredient of concern is the ingredient present in the greatest amount in that feedstuff, or the reason a particular feed is being used. For a *protein supplement*, this would be protein. For a *mineral* mix, it might be either calcium or phosphorus. However, since the cost of phosphorus is generally about ten times greater than the cost of calcium, phosphorus is the major ingredient of economic concern in most mineral supplements. For *cereal grains*, and generally for hay, the major ingredient of concern is *energy*, since 80 to 90% of the feed ingested is necessary to provide energy for the animal.

To determine the cost of the major ingredient of concern in a feedstuff, divide the cost of the feed by the amount of that ingredient present in the feed, as shown in the following examples. As previously discussed in this chapter, it is important that comparisons be made among feeds with an equal moisture content. To ensure that this is done, it is best to convert the amount of nutrients in the feed to the amount present in the feed *dry matter.*

Example 1

The following two hays are available, and will be fed to mature, nonpregnant, nonlactating horses. Which hay is more economical to feed?

Hay	Cost/ton of hay ($)	kcal/lb	Protein (%)
Alfalfa	60	1000	17
Grass	55	820	8

Both hays contain enough *protein* to meet the horse's requirements (8%). Therefore, the difference in their protein content is not of concern. The ingredient of major concern in the hays is the *energy* they provide, so we compare their cost on an energy basis.

Alfalfa: $60 ÷ 1000 = $0.060/kcal
Grass: $55 ÷ 820 = $0.067/kcal

The alfalfa costs less per unit of energy, and therefore is a better buy than the grass hay. The *energy* provided by a feed may be expressed in *kilocalories* as in this example, or in the percentage of *total digestible nutrients (% TDN).* For example, if instead of kilocalories, the energy content was given as 50% TDN in the alfalfa and 40% in the grass hay, the calculations to compare their cost would be as follows.

Alfalfa: $60 ÷ 0.50 = $120.00/ton of TDN
Grass: $55 ÷ 0.40 = $137.50/ton of TDN

Example 2

If the following *mineral* mixes are available, which is the most economical?

Mineral	Cost/cwt ($)	Ca (%)	P (%)
Monosodium phosphate	16.25	0	24
Dicalcium phosphate	13.00	24	19
Limestone	2.30	35	0

Monosodium phosphate: $16.25 ÷ 0.24 = $67.71/cwt of P
Dicalcium phosphate: $13.00 ÷ 0.19 = $68.42/cwt of P
Limestone: $ 2.30 ÷ 0.35 = $6.57/cwt of Ca

Thus, if phosphorus was the only *mineral* of concern, monosodium phosphate would be the most economical. If both calcium and phosphorus are needed, the cost of the calcium in dicalcium phosphate must be added to the cost of monosodium phosphate. Since 100 lbs of dicalcium phosphate provides 24 lbs of calcium, the cost of 24 lbs of calcium present in the dicalcium phosphate must be added to the cost of the monosodium phosphate. The cost of 24 lbs of calcium is ($6.57/100 lbs) × (24 lbs) = $1.58. Thus, 100 lbs of monosodium phosphate plus 24 lbs of limestone provides the same amount of phosphorus and calcium as 100 lbs of dicalcium phosphate. The cost of the calcium and phosphorus in the monosodium phosphate plus limestone mixture is $67.71 + $1.58 = $69.29, as compared with a cost of $68.42 for dicalcium phosphate. Therefore, if both phosphorus and calcium are needed, dicalcium phosphate is the most economical.

Example 3

The following two types of hay are available. Which hay is more economical for feeding yearlings?

Hay	TDN (%)	Protein (%)	Cost/ton of hay
Alfalfa	55	17	$62
Grass	45	8	$45

If only the cost of *energy* is considered, the grass hay would be more economical.

$$\text{Alfalfa:} \quad \$62 \div 0.55 = \$112.73/\text{ton of TDN}$$
$$\text{Grass:} \quad \$45 \div 0.45 = \$100.00/\text{ton of TDN}$$

The yearling's *ration* needed to provide optimum growth rate should consist of a maximum of one-half *concentrate* and a minimum of one-half *roughage*. *Cereal grains* contain about 12% *protein* (Appendix Table 1), and yearlings need 12% protein in their ration (Appendix Table 2). If alfalfa is fed, the yearling's ration will contain ½ (12%) + ½ (17%), or 14.5% protein, so a protein *supplement* is not needed. If grass hay is fed, the ration will contain ½ (12%) + ½ (8%), or 10% protein. Therefore, a protein supplement such as soybean meal must be added to the grain. Both the *energy* and protein provided by the two types of hay must be considered in comparing their relative values. This can be done by inserting the cost, energy, and protein content of the feeds available into the following formulas:

$$y = \frac{(\% \text{ protein in feed evaluated}) (2000) - \dfrac{(\% \text{ protein in grain}) (\% \text{ TDN in feed evaluated}) (2000)}{(\% \text{ TDN in grain})}}{(\% \text{ protein in protein supplement}) - \dfrac{(\% \text{ TDN in protein supplement}) (\% \text{ protein in grain})}{(\% \text{ TDN in grain})}}$$

$$x = \frac{(\% \text{ TDN in feed evaluated}) (2000) - (\% \text{ TDN in protein supplement}) (y)}{(\% \text{ TDN in grain})}$$

$ value of feed being evaluated/ton = (x) ($/lb of grain) + (y) ($/lb of protein supplement)

If the value of the feed being evaluated is greater than its purchase price, it is a good buy. In contrast, if its value is less than its cost, it is not a good buy with respect to the *cereal grain* and *protein supplement* with which it was compared. Although these formulas may look formidable, they are easy to use. Simply insert the values for the feeds, and do the arithmetic involved, as shown for this example.

In this example, let's say that the *grain* and *protein supplements* available are as follows:

Feed	Cost/ton ($)	TDN (%)	Protein (%)
Oats	110	68	12
Corn	100	80	10
Soybean meal	190	74	44
Cottonseed meal	180	68	41

First, it is necessary to determine which of these *concentrates* to use. If the decision is based only on cost (for other considerations see the discussion of cereal grains in this Chapter), corn and soybean meal would be used. Since in this example, corn is higher in *energy* and lower in cost than any of the other concentrates, it is obviously the most economical source of energy, and soybean meal is the most economical source of *protein* as shown below.

Feed	Cost/ton ($)		Protein (%)		Cost/ton of Protein ($)
Oats	110	÷	0.12	=	917
Corn	100	÷	0.10	=	1000
Soybean meal	190	÷	0.44	=	432
Cottonseed meal	180	÷	0.41	=	439

Using corn and soybean meal as the *concentrates* in the formulas, it can be determined which hay is the best buy.
Alfalfa:

$$y = \cfrac{(.17)\,(2000) - \cfrac{(.10)\,(.55)\,(2000)}{(.80)}}{(.44) \quad - \quad \cfrac{(.74)\,(.10)}{.80}} = 583$$

$$x = \frac{(.55)\,(2000) - (.74)\,(583)}{(.80)} = 836$$

$$\$ \text{ value of alfalfa} = (836)\frac{\$100}{2000} + (583)\frac{\$190}{2000} = \$97.18/\text{ton}$$

The value of the alfalfa hay with respect to its cost = $97.18 ÷ $62.00 = 1.57. This means that the alfalfa is worth 1.57 times more than the cost of an equal amount of *protein* and *energy* provided by corn and soybean meal.
Grass:

$$y = \cfrac{(.08)\,(2000) - \cfrac{(.10)\,(.45).(2000)}{(.80)}}{(.44) \quad - \quad \cfrac{(.74)\,(.10)}{(.80)}} = 137$$

$$x = \frac{(.45)\,(2000) - (.74)\,(137)}{(.80)} = 998$$

$$\$ \text{ value of grass hay} = (998)\frac{\$100}{2000} + (137)\frac{\$190}{2000} = \$62.92/\text{ton}$$

The value of the grass hay with respect to its cost = $62.92 ÷ $45.00 = 1.40

With respect to both the *protein* and the *energy* in the hay, and the cost of corn and soybean meal, alfalfa is worth $97.18/ton, which is 1.57 times more than its cost. The grass hay is worth $62.92/ton, which is 1.40 times its cost. Since the value of the alfalfa with respect to its cost is more than the value of the grass hay with respect to its cost (1.57× *vs* 1.40×), alfalfa is more economical.

Example 4

The following two *cereal grains* are available. Which grain is more economical to use in a *concentrate* mix for weanlings being fed grass hay?

Grain	TDN (%)	Protein (%)	Cost/100 lbs
Barley	75	12	$5.20
Corn	80	10	$5.30

If only the cost of *energy* is considered, corn is more economical.

> Barley: $5.20 ÷ 0.75 = $6.93/cwt of TDN
> Corn: $5.30 ÷ 0.80 = $6.63/cwt of TDN

In contrast, if only the cost of *protein* is considered, barley is more economical.

> Barley: $5.20 ÷ 0.12 = $43.33/cwt of protein
> Corn: $5.30 ÷ 0.10 = $53.00/cwt of protein

The weanling's *ration* needed to provide optimum growth rate should consist of a maximum of one-half *concentrate* and a minimum of one-half *roughage*. Thus the amount of *protein* present in the ration will be ½ times the % protein in the concentrate mix plus ½ times the % protein in the roughage. Grass hay contains about 8% protein (Appendix Table 1), and weanlings need 14.5% protein in their ration (Appendix Table 2). If barley is the concentrate fed, the weanling's ration will contain ½ (12%) + ½ (8%), or 10% protein. If corn is the concentrate fed, the weanling's ration will contain ½ (10%) + ½ (8%), or 9% protein. Thus, regardless of which *cereal grain* is used, a protein *supplement* is needed in the concentrate mix. However, less protein supplement would be needed if barley rather than corn were used. Thus, both the *energy* and the protein provided by the cereal grains must be considered in comparing their relative values. This can be done by inserting the cost, energy, and protein content of the feeds available into the formulas given in Example 3. In this example, let's say that the soybean meal, as given in Example 3, is available, and is the most economical source of protein. Therefore, it will be used to compare the cereal grains too. Either cereal grain may be used in the formulas as the feed being evaluated and the other one used as the grain with which it is being compared. Let's use barley as the feed being evaluated.

$$y = \dfrac{(\% \text{ protein in barley}) (2000) - \dfrac{(\% \text{ protein in corn}) (\% \text{ TDN in barley}) (2000)}{(\% \text{ TDN in corn})}}{(\% \text{ protein in soybean meal}) - \dfrac{(\% \text{ protein in corn}) (\% \text{ TDN in soybean meal})}{(\% \text{ TDN in corn})}}$$

$$x = \dfrac{(\% \text{ TDN in barley}) (2000) - (\% \text{ TDN in soybean meal}) (y)}{(\% \text{ TDN in corn})}$$

$$y = \dfrac{(.12) (2000) - \dfrac{(.10) (.75) (2000)}{(.80)}}{(.44) - \dfrac{(.74) (.10)}{(.80)}} = 151$$

$$x = \dfrac{(.75) (2000) - (.74) (151)}{(.80)} = 1735$$

$$\$ \text{ value of barley/ton} = (1735 \text{ lbs}) \left(\dfrac{\$5.30}{100 \text{ lbs of corn}} \right) + (151 \text{ lbs}) \left(\dfrac{\$190}{2000 \text{ lbs of soybean meal}} \right)$$

$$= \$91.96 + \$14.35 = \$106.31/\text{ton}$$

$$= \left(\dfrac{\$106.31}{\text{ton}} \right) \left(\dfrac{1 \text{ ton}}{2000 \text{ lbs}} \right) (100 \text{ lbs}) = \$5.32/100 \text{ lbs}$$

Thus, in regard to both *protein* and *energy*, if soybean meal cost $190/ton and corn $5.30/100 lbs, barley is worth $5.32/100 lbs, yet it costs only $5.20/100 lbs. Since barley is worth more than its cost as compared with corn, it is more economical than corn. In contrast, if its value was found to be less than its cost with respect to the *cereal grain* it was being compared with, then the other cereal grain would be more economical.

Chapter 3
Ration Formulation and Evaluation

Formulating a *ration** or evaluating a ration that is to be fed and correcting it to ensure that it meets the nutritional requirements of the animal is quite easy once the procedure is learned. Although it may seem difficult at first, the same procedure is used to determine the amount of any *nutrient* in the ration, or to determine how much of a feed containing that nutrient needs to be added to the ration to meet the animal's requirements. This procedure is then used for each nutrient.

The first thing necessary to formulate or evaluate a *ration* is to know the *nutrient* content of the feeds available or being used. The values given in Appendix Table 1 are generally quite adequate for *concentrates*, and may be used for *forages*. However, since the nutrient content of different forages may vary considerably, it is best to have the specific forage analyzed for its nutrient content. How to determine the amount of nutrients present in feeds is described in Chapter 2. Once these values are known, the nutrients in the ration should be evaluated or formulated in the order of the amount of each needed by the animal. This order is first *energy*, next *protein*, then calcium, and phosphorus. If average or better quality feeds are being used, these are the only nutrients that need to be evaluated in the ration. The amounts of other nutrients in the ration may be evaluated, or formulated, to ensure that proper amounts are present, in the same manner as described for energy, proteins, calcium, and phosphorus. However, always ensure that the ration first meets the requirements for these four nutrients before considering others.

Since over 80% of the *ration* eaten by the animal is needed for *energy*, the amount of energy available is the first consideration. This condition is met by simply ensuring that there are adequate quantities of feed available to supply the energy needed to maintain the animal

---

*Words in italics are defined in the glossary.

at optimum body weight and condition, or to achieve optimal growth rate. If the animal cannot consume enough of the feed to meet its energy needs, a higher energy density feed, such as *cereal grains,* must be fed. As a general rule, *concentrates* should never make up more than one half of the total weight of feed eaten by the horse. The total amount of feed that the horse will eat is given in Appendix Table 2. The maximum amount of concentrates that should be fed to most growing horses to obtain optimal growth rate, with a minimal chance of *epiphysitis,* crooked legs, and contracted *flexor tendons,* is given in Table 8–3.

CALCULATING THE AMOUNT OF FEED NEEDED

The amount of feed needed can be calculated as shown in the following examples. Remember, these are examples only. Several factors must be considered in determining the correct amount to feed the horse. These factors are (1) variation among horses in their *energy* needs, (2) the amount of feed eaten and the amount wasted, (3) weather conditions, and (4) energy content of feeds. **Always feed the amount needed to maintain the horse at optimum body weight and condition.**

Example 1

How much feed does a 900-lb mature horse need when not working? As given in Appendix Table 2, the 1100-lb mature horse, at rest, needs 8.2 lbs of *TDN*/day; for each 100 lbs above or below this, add or subtract 8%. Therefore, the 900-lb horse would need (1100 − 900) ÷ 100 or 2 × 8% less than the 1100-lb horse. The amount needed therefore would be

$$[(100\%) - (2) \times (8\%)] \times (8.2) = (84\%) \times (8.2) = 6.9 \text{ lbs TDN/day.}$$

The amount of feed necessary to provide this amount of *energy* can be determined by dividing the amount of energy needed by the energy content of the feed (Appendix Table 1). Thus, if early bloom alfalfa hay, which contains 50% TDN, were the only feed being used, the horse would need to consume 6.9 lbs of TDN/day ÷ 50% TDN = 13.8 lbs of hay/day. Generally, an additional 10 to 15% of feed must be added to allow for that wasted and lost. If 2 lbs of oats containing 68% TDN were fed, they would provide (2 lbs) × (68% TDN) or 1.36 lbs of TDN. The amount of hay needed would be (6.9 − 1.36) ÷ 50% TDN in the hay or 11 lbs per day, plus an additional 1 to 2 lbs to allow for loss.

Example 2

A 1400-lb horse used at a slow trot for 3 hours would need:
Energy needed for rest = [100% + (3) × (8%)] × (16.4 Mcal/day) =
20.3 Mcal/day (see Example 1 for explanation of calculation)
Energy needed for work = (body wt) × (hrs of activity) × (energy
needed from Appendix Table 3), which in this example = (1400
lbs) × (3 hrs) × (0.23 Mcal/hr/100 lbs) = 9.7 Mcal.
Total energy needed = 20.3 + 9.7 = 30 Mcal/day
Timothy hay needed = (30 Mcal/day) ÷ (1.8 Mcal/kg in timothy
hay, Appendix Table 1) = 17 kg or 37 lbs/day
This equals 2.6% of the horse's body weight per day (37 lbs ÷ 1400
= 2.6%).
Since the mature horse can eat 3% of its body weight per day, this
amount could be consumed.
If the horse is fed hay for maintenance, and we supply the additional *energy* needed by feeding corn, the following would apply:
Timothy hay needed at rest: (20.3 Mcal/day) ÷ (1.8 Mcal/kg in
timothy hay, Appendix Table 1) = 11.3 kg or 25 lbs/day
Corn needed for work: (9.7 Mcal) ÷ (3.5 Mcal/kg in corn, Appendix
Table 1) = 2.77 kg or 6 lbs or (6 lbs corn) ÷ (1.7 lbs/qt, Table 2–2
or Appendix Table 8) = 3.5 qts of corn
If the horse was used as described in this example twice a week, 7
qts of corn would be needed weekly. Therefore, the feeding program
would be 25 lbs of hay, plus 2 to 3 lbs to allow for loss, and 1 qt of corn
per day.

Example 3

An 850-lb horse to be used 4 hours on an endurance ride at a fast trot
and canter would need:
Energy needed for rest = [100% − (2.5) × (8%)] × (8.2 lb TDN /day)
= 6.56 lb TDN/day (see Example 1 for explanation of calculation)
Energy needed for work = (body wt) × (hrs of activity) × (energy
needed from Appendix Table 3) which in this example = (850 lbs)
× (4 hours) × (0.29 lbs TDN/hr/100 lbs) = 9.86 lbs of TDN/day
Total energy needed = 6.56 + 9.86 = 16.42 lbs of TDN
Early bloom alfalfa hay needed = 16.42 ÷ (50% TDN in alfalfa,
Appendix Table 1) = 33 lbs
This amount would be 3.9% of the horse's body weight (33 ÷ 850 =
3.9%), which is more than can be eaten daily (Appendix Table 2). It
would be necessary to feed some *grain* to provide the remaining
amount of *energy* needed. Grain should not exceed one half of the
horse's *ration* by weight. To determine the amount of grain and

roughage this would be, divide the amount of energy needed by the sum of their energy contents. In this example let's use corn as the cereal grain fed. The amount of TDN needed is 16.42 lbs/day. Corn contains 80% TDN, and the roughage (alfalfa) contains 50% TDN (Appendix Table 1). Therefore, the maximum amount of corn that should be fed is (16.42) ÷ (0.80 + 0.50) = 12.6 lbs/day. The alfalfa should be available for the horse to eat as much as it will consume without wastage, which would be about 12.6 lbs daily. As shown in the following, 12.6 lbs per day each of corn and alfalfa would provide the 16.42 lbs of TDN needed by the horse daily.

Alfalfa	(12.6 lbs) × (50% TDN) =	6.3 lbs of TDN
Corn	(12.6 lbs) × (80% TDN) =	10.1 lbs of TDN
Total	25.2 lbs of feed and	16.4 lbs of TDN

Thus the total amount of feed per day that is being fed is equal to 3% of the horse's body weight (25.2 ÷ 850). Since the horse is capable of eating an amount equal to 3% of its body weight daily, this amount could be consumed. **Although the use of the horse each day may vary, the amount of feed received daily should remain fairly constant.** Thus, it is best to calculate the horse's *energy* needs over a week or so, and feed the average amount needed daily as described in the following example.

Example 4

The 850-lb horse in the preceding example is worked at a slow trot for one-half hour, then at a faster trot, and a canter for 2 hours daily, 3 days per week. How much timothy hay, which contains 40% *TDN*, and oats, containing 68% TDN (Appendix Table 1), should be fed to meet the animals's *energy* requirements? A feeding program that works well is to feed a sufficient amount of hay to supply energy needs for maintenance, and enough *grain* to supply the additional energy needed for physical activity. Thus, the amount of hay fed is always the same. The amount of grain may be increased or decreased as needed; however, it is best to feed a fairly constant amount of grain each day also. In this example, the following feeding program could be used: Hay needed for rest = 6.56 lbs TDN/day (from Example 3) ÷ 40% TDN = 16.4 lbs of hay/day.

Energy needed for work = (850 lbs) × (½ hour) × (0.12 lbs TDN/hour/100/lbs body wt) (from Appendix Table 3) + (850 lbs) × (2 hours × (0.29 lb TDN/hour/100 lbs body wt) (from Appendix Table 3) = 0.51 + 4.93 = 5.44 lbs TDN/day.
(3 days/week) × (5.44 lbs TDN/day) = 16.32 lbs TDN/week or an average of 16.32 ÷ 7 or 2.33 lbs of TDN/day.

Oats needed = 2.33 lbs ÷ 68% TDN in oats = 3.4 lbs of oats/day. The recommended feeding program is 16 to 17 lbs of hay, plus 2 to 3 lbs to allow for loss, and 3 to 4 lbs of oats per day.

When the horse is used in an endurance race, such as in Example 3, additional *energy* is needed. In Example 3, it was found that 9.86 lbs of *TDN* were needed for the endurance race described. This would require feeding 14.5 lbs of oats (9.86 lbs of TDN ÷ 68% TDN in oats). Feeding this amount of oats all in one day to a horse unaccustomed to this much may cause *founder* or *colic*, and should not be done. The horse will consume additional amounts of *roughage*, if it is available, in order to obtain the additional energy needed. Therefore, all the roughage that the horse wants to eat should be available for at least several days after hard physical work. Additional energy should also be provided by increasing the amount of *grain* fed for several days. In this example, all the hay the horse wants to eat should be available after the endurance race, and the amount of oats fed should be increased from 3 to 4 lbs per day to 5 to 7 lbs per day for the next two to three days. **The best feeding program is to feed as much hay as the horse will eat without wastage, and as much grain as needed to keep the horse at optimum body weight and condition.**

FORMULATING A COMPLETE RATION AND EVALUATING ITS NUTRITIVE CONTENT

The *ration* should be formulated, or evaluated, for each *nutrient* in the order of the amount needed by the animal. First ensure that there is adequate feed available to provide the *energy* necessary to maintain the animal at optimum body weight and condition, or to achieve optimal growth rate. After this is done, ensure that the ration contains adequate *protein*, then calcium, and last phosphorus, to meet the animal's requirement for these nutrients (Appendix Table 2). If the horse has water and *trace-mineralized* salt available, and is receiving average or better quality feeds, the amount of feed, and its protein, calcium, and phosphorus content, are the only things that need to be checked in the ration. The same procedure described may be used to formulate, or evaluate, a ration for any nutrient. It is suggested that the reader do the mathematics for each of the examples given, to become familiar with the procedure.

To formulate a *ration*, the *nutrients* present in the *roughage* being fed, as determined from Appendix Table 1, or preferably by laboratory analysis, should be compared to the animal's requirements for these nutrients. A *concentrate* mix containing the amount of nutrients necessary to make up any deficiencies may be formulated as described, or in many cases, a commercially prepared ration may be purchased.

Having *grains* mixed with *protein,* calcium, and phosphorus supplements (or supplements to provide any *nutrients* needed to meet the horse's requirements) at the feed mill has several advantages, as well as disadvantages. The advantages are as follows: (1) feeding is easier and faster; (2) there is less chance of error at feeding, particularly when several supplements are needed; (3) it is generally less expensive than buying commercially prepared mixes; and (4) the exact amount needed of each nutrient can be included in the grain mix, so with the *roughage* being fed, it will meet the particular horse's requirements.

The major disadvantages of having a *grain ration* mixed at the feed mill are as follows: (1) large quantities must be purchased (usually several tons), and (2) errors and poor mixing of the ration, unfortunately, are not uncommon. For this reason, at least two samples from different parts of the mix should be analyzed to see that it contains the correct amount of *nutrients* (primarily *protein,* calcium, and phosphorus). The feed mill should be informed **before** the grain ration is mixed that it will be analyzed and cannot be used if it is not what is wanted. The local agricultural extension service or veterinarian usually has the information on where feeds can be analyzed. The necessity of having the grain ration analyzed increases its cost, which, in addition to the large amount that must be mixed, makes this an impractical and uneconomical method for those needing enough for only a few horses. However, it may be much less expensive than buying commercially prepared grain mixes for those feeding a larger number of horses.

Example 1

A 600-lb weanling is on grass pasture. The grass is primarily mature timothy and brome, with ample quantities for the weanling to eat all it wants. Formulate a *grain* mix that, when fed in amounts necessary to obtain optimal growth rate, will, along with the pasture, meet the weanling's *protein,* calcium, and phosphorus requirements.

Six steps are necessary to do this.

1. Determine the amounts of concentrate and roughage that will be eaten, and the fraction of each in the ration. The amount of concentrate to feed for optimum growth rate is given in Table 8–3, and the total amount of feed the horse will consume is given in Appendix Table 2.

2. Determine the nutrients present in the roughage as given in Appendix Table 1, or more accurately, by having it analyzed (see Chapter 2).

3. Determine the amount of each nutrient that the roughage contributes to the ration. This is done by multiplying the amount of each

nutrient in the roughage by the fraction of the ration that is roughage (% nutrient in roughage × fraction of roughage in ration = % of that nutrient the roughage contributes to the ration).

4. Determine the amount of each nutrient that the concentrate mix must contribute to the ration to supply that needed by the animal. This is done by subtracting the amount of each nutrient supplied by the roughage, from the total amount of that nutrient needed by the animal (% of that nutrient required in the ration to meet the horse's needs as given in Appendix Table 2 and Table 1–5 – % of that nutrient the roughage contributes to the ration as determined in step 3 = % of that nutrient that the concentrate mix must contribute to the ration).

5. Determine the amount of each nutrient that must be present in the concentrate mix to meet the animal's requirements. This is done by dividing the amount of each nutrient that the concentrate mix must contribute to the ration (the values determined in step 4) by the fraction of the ration that is concentrate (the value determined in step 1).

6. Buy a commercial concentrate mix, or formulate a concentrate mix, that contains at least the amount of each nutrient needed (the values determined in step 5).

The calculations for these steps in this example are as follows:

1. The amount of *concentrate* and *roughage* in the *ration:* The total amount of feed the weanling will eat is equal to 3% of its body weight daily (Appendix Table 2)—in this case, 18 lbs/day (600 lbs body wt × 3%). To obtain optimum growth rate, the maximum amount of concentrate that should be fed is 1.5 lbs/100 lbs body wt/day (Table 8–3)—in this case, 9 lbs/day (1.5 lbs/100 lbs body wt/day × 600 lbs body wt). This would leave room for the weanling to eat 9 lbs of pasture *forage* daily (18 lbs total feed – 9 lbs of concentrate). Thus, the fraction of the total ration that is concentrates is 0.5 (9 lbs concentrate ÷ 18 lbs total feed), and the fraction that is roughage is 0.5 (9 lbs of roughage ÷ 18 lbs total feed).

2. The amounts of *nutrients* most likely present in the *roughage* are: 6% *protein*, 0.35% calcium, and 0.15% phosphorus (Appendix Table 1).

3. The amount of each *nutrient* that the *roughage* contributes to the *ration:*

Nutrient	Fraction of roughage in the ration (from step 1)		% nutrient in feed (from step 2)		% nutrient in ration from *roughage*
protein	0.5	×	6%	=	3.0%
calcium	0.5	×	0.35%	=	0.175%
phosphorus	0.5	×	0.15%	=	0.075%

4. The amount of each *nutrient* that the *concentrate* mix must contribute to the *ration:*

Nutrient	% needed by animal (from Appendix Table 2)		% nutrient in ration from roughage (from step 3)		% nutrient needed in ration from concentrate
protein	14.5	−	3.0	=	11.5
calcium	0.65	−	0.175	=	0.475
phosphorus	0.45	−	0.075	=	0.375

5. The amount of each *nutrient* that must be in the *concentrate:*

Nutrient	% nutrient needed in ration from concentrate (from step 4)		Fraction of conc. in the ration (from step 1)		% nutrient needed in concentrate
protein	11.5	÷	0.5	=	23
calcium	0.475	÷	0.5	=	0.95
phosphorus	0.375	÷	0.5	=	0.75

This entire procedure (all five steps) can be put into a table and calculated as shown in Table 3–1.

6. To meet the weanling's requirements when it is on the described grass pasture, it must be fed 9 lbs per day of a *concentrate* mix containing 23% *protein*, 0.95% calcium, and 0.75% phosphorus. A commercial mix containing these amounts of *nutrients* could be purchased, or a concentrate mix could be formulated.

The *cereal grains* contain 10 to 12% *protein*. For this discussion let's use oats as an example. Oats contain 12% protein (Appendix Table 1), therefore, enough protein *supplement* must be added to the oats to increase the protein content to 23%. As discussed in Chapter 1, soybean meal (SBM) is the plant-source *protein supplement* preferred for the growing horse. From Appendix Table 1, we find that SBM contains 44% protein. The amount of SBM needed may be determined using the method known as the Pearson square.

TABLE 3–1
CALCULATING THE AMOUNT OF NUTRIENTS NEEDED IN THE CONCENTRATE MIX

Feed	Pounds eaten daily	Fraction of the ration†	Protein (%) in feed	Protein (%) in ration	Calcium (%) in feed	Calcium (%) in ration	Phosphorus (%) in feed	Phosphorus (%) in ration
Roughage	9*	0.5	6‡	3.0§	0.35‡	0.175§	0.15‡	0.075§
Concentrate	9	0.5	23**	11.5#	0.95**	0.475#	0.75**	0.375#
Total‖	18	1.0		14.5‖		0.650‖		0.450‖

*The amount of roughage consumed = total amount of feed eaten (Appendix Table 2) minus the amount of concentrate fed.

†Fraction of the ration = the amount of each feed eaten divided by the total amount eaten (9 ÷ 18 = 0.5).

‡From Appendix Table 1, or more accurately by analysis.

§The fraction of the ration consisting of *roughage* times the amount of each *nutrient* present in the roughage (0.5 × 6 = 3.0, 0.5 × 0.35 = 0.175 and 0.5 × 0.15 = 0.075).

‖The total amount of each nutrient in the ration must be equal to that needed by the animal, as given in Appendix Table 2.

#The amount of each nutrient provided to the ration by the concentrate = the total amount needed minus the amount provided by the roughage (14.5 − 3.0 = 11.5, 0.65 − 0.175 =0.475, and 0.45 − 0.075 = 0.375).

**The amount of each nutrient needed in the concentrate = the amount of each nutrient that must be provided to the ration by the concentrate divided by the fraction of the ration consisting of concentrate (11.5 ÷ 0.5, = 23, 0.475 ÷ 0.50 = 0.95 and 0.375 ÷ 0.50 = 0.75).

PEARSON SQUARE

The Pearson square method for determining the amount of a feed needed in the ration is conducted in the following manner. As shown on the following page, the percent of the specific *nutrient* required is placed in the center of the square, in this example, 23%. The percentage of that nutrient in the *ration* (or in this case, in the *grain*) is placed at the upper left corner. In this case, it is 12%. The percent of that nutrient in the *supplement* is placed at the lower left corner. Here it is 44%. Next, the number in the middle is subtracted from the number at the lower left corner, and the answer is placed at the upper right corner: in this case 44 − 23 = 21. Then the number at the upper left corner is subtracted from the number in the middle, and the answer is placed at the lower right corner: in this case 23 − 12 = 11. Add the two numbers at the right corners: in this case 21 + 11 = 32. Then divide the number at the lower right corner by this sum: in this case 11 ÷ 32 = 0.344. The value obtained is the fraction of the supplement needed in the ration, in this case SBM with the oats, to increase the nutrient content of the mixture to the amount required.

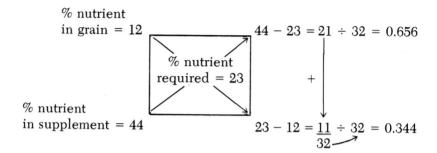

% nutrient
in grain = 12   44 − 23 = 21 ÷ 32 = 0.656

% nutrient
required = 23   +

% nutrient
in supplement = 44   23 − 12 = 11 ÷ 32 = 0.344

As determined using the Pearson square, in order for a mixture of oats and SBM to contain 23% *protein*, there must be 34.4% SBM and 65.6% oats (100 − 34.4, or 21 ÷ 32).

Oats contains 0.1% calcium and 0.3% phosphorus, and SBM contains 0.25% calcium and 0.6% phosphorus (Appendix Table 1). The *concentrate* mix must contain 0.95% calcium and 0.75% phosphorus (Table 3–1), therefore a calcium-phosphorus *supplement* is needed. The *concentrate* mix must be pelleted or liquid molasses must also be added to hold it together and keep the SBM and *mineral* supplement from sifting out. Usually, 3 to 10% molasses works well (see Chapter 2). Since this *ration* contains so much SBM, the higher amount of molasses is needed. If this isn't sufficient to prevent the SBM and mineral from sifting out, the ration should be pelleted.

The *mineral supplement* doesn't contain any *protein*, and the

TABLE 3–2
CALCULATING THE AMOUNT OF CALCIUM AND PHOSPHORUS IN THE CONCENTRATE MIX

Feed	Fraction of the conc. mix	Calcium (%) in feed*	Calcium (%) in ration†	Phosphorus (%) in feed*	Phosphorus (%) in ration†
Oats	0.52	0.10	0.052	0.30	0.156
SBM	0.38	0.25	0.095	0.60	0.228
Molasses	0.10	0.90	0.090	0.15	0.015
TOTAL	1.00		0.237		0.399
Required in concentrate mix (Table 3–1)			0.95		0.75

*From Appendix Table 1, or more accurately, by analysis.

†Amount of each nutrient that is contributed to the total ration, or mix, by that feed equals the fraction of that feed in the ration (or mix) times the amount of the nutrient in that feed, e.g., the fraction of oats in the ration is 0.52, and oats contains 0.10% calcium; therefore, oats contributes (0.52) × (0.10%) or 0.052% calcium to the ration.

protein in molasses is low, so when they are added to the *concentrate* mix it will lower the total protein content of the mixture. To compensate for this, 3 to 4% protein supplement above that previously calculated should be added. Thus, instead of 34.4% SBM, let's use 38%. The concentrate mix in this example would then contain 38% SBM, 10% molasses, 52% oats, and a calcium-phosphorus supplement. To determine how much mineral supplement is needed, we must first determine how much calcium and phosphorus are present in the concentrate mix without it. This is done by multiplying the fraction of each feed in the concentrate mix times its calcium and phosphorus content, and then summing the amount of each mineral contributed by each feed as shown in Table 3–2.

Once it is known how much calcium and phosphorus there is in the *concentrate* mix and the amount of calcium and phosphorus present in the *mineral supplement* used, the amount of that supplement needed can be determined using the Pearson square (see earlier in this example for an explanation of this procedure). Since both calcium and phosphorus are needed, let's use dicalcium phosphate (dical), which contains 24% calcium and 19% phosphorus (Appendix Table 5). Since the concentrate mix in this example is more deficient in calcium than in phosphorus, determine how much dical is needed to provide adequate calcium, rather than how much is needed to provide adequate phosphorus.

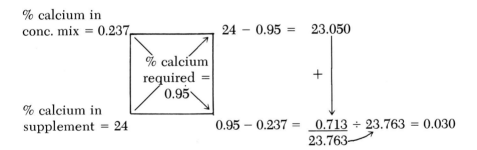

As determined using the Pearson square, in order to increase the calcium content of the *concentrate* mix to 0.95%, add 3% dicalcium phosphate. Decrease the amount of oats in the mixture by this amount. Next, as shown in Table 3–3, check to see if the *concentrate* mix now contains the *nutrient* levels required as determined in Table 3–1.

The entire *ration* should be checked as shown in Table 3–4 to ensure that the weanling's nutritional requirements are met when this *concentrate* mix is fed along with all the pasture grass that can be consumed.

TABLE 3–3
CALCULATING THE NUTRIENT CONTENT OF THE CONCENTRATE MIX

Feed	Fraction of feed in concentrate mix	Protein (%) in feed*	Protein (%) in ration†	Calcium (%) in feed*	Calcium (%) in ration†	Phosphorus (%) in feed*	Phosphorus (%) in ration†
Oats	0.49	12	5.88	0.10	0.049	0.30	0.147
SBM	0.38	44	16.72	0.25	0.095	0.60	0.228
Molasses	0.10	4	0.40	0.90	0.090	0.15	0.015
Dical	0.03	0	0.00	24.0	0.720	19.0	0.570
TOTAL	1.00		23.00		0.954		0.960
Required in concentrate mix as determined in Table 3–1			23		0.95		0.75

*,†See footnotes in Table 3–2 for explanation of how these values were obtained.

TABLE 3–4
CALCULATING THE AMOUNT OF NUTRIENTS PRESENT IN THE TOTAL RATION

Feed	Pounds eaten daily	Fraction of the ration†	Protein (%) in feed	Protein (%) in ration	Calcium (%) in feed	Calcium (%) in ration	Phosphorus (%) in feed	Phosphorus (%) in ration
Roughage	9*	0.5	6‡	3.0§	0.35‡	0.175§	0.15‡	0.075§
Conc. mix	9	0.5	23‖	11.5#	0.954‖	0.477#	0.960‖	0.480#
TOTAL	18	1.0		14.5**		0.652**		0.555**
Requirements for the weanling (from Appendix Table 2)				14.5		0.65		0.45

*,†,‡,§See footnotes Table 3–1 for explanation of how these values were obtained.
‖Total nutrients present in the concentrate mix as determined in Table 3–3.
#The fraction of the ration consisting of the concentrate mix times the amount of each nutrient present in the concentrate mix ($0.5 \times 23 = 11.5$, $0.5 \times 0.954 = 0.477$, and $0.5 \times 0.960 = 0.480$).
**Total nutrients present in the ration equals the sum of the nutrients provided by each ingredient in the ration, in this case roughage and concentrate mix. If this value is equal to or greater than that required by the animal, the ration meets the animal's requirements for these nutrients.

Finally, check the *Ca:P ratio* in the *ration* to ensure that it is in the acceptable range of 1:1 to 3:1, by dividing both the amount of calcium (0.652%) and the amount of phosphorus (0.555%) present in the total ration by the amount of phosphorus (0.555%).

$$Ca:P = 0.652:0.555 = \frac{0.652}{0.555} : \frac{0.555}{0.555} = 1.17:1$$

Instead of having the *concentrate* mixed at a feed mill and then feeding 9 lbs/day of the concentrate mixture, the ingredients could be mixed at each feeding. If this were done, the following amounts would be needed daily, as determined by multiplying the fraction of each ingredient in the concentrate mix (Table 3–3) by the amount of concentrate mix fed. However, since the concentrate mix will not contain molasses when fed in this manner, add the fraction of the mix that was molasses to the grain; in this case oats.

Oats: 0.59×9 lbs/day = 5.3 lbs/day
SBM: 0.38×9 lbs/day = 3.4 lbs/day
Dical: 0.03×9 lbs/day = 0.27 lbs/day $\times$ (16 oz/lb) = 4.3 oz/day

If the grain does not contain molasses, it is necessary to dampen it with water at feeding to prevent the SBM and dical from sifting out. If the feed is dampened, and all of it is not eaten within a few hours, any remaining feed should be removed and discarded. Dampened feed will become moldy rapidly. Moldy feed is unpalatable, and may be toxic to the horse.

Example 2

A 500-lb weanling is being fed 6 lbs of oats and all the good quality, early-bloom alfalfa hay it will consume. To determine the *nutrients* present in this *ration*, the amount of alfalfa eaten must be known. The maximum amount of feed that the weanling can eat daily is about 3 lbs per 100 lbs of body weight (Appendix Table 2). Thus, this weanling will eat 15 lbs of feed per day. Since 6 lbs of oats is being fed, 9 lbs (15 − 6) of alfalfa is consumed.

The nutritive content of each feedstuff in the *ration* may be obtained from Appendix Table 1, or more accurately, by analysis. The fraction of each feedstuff in the total ration is then multiplied by the *nutrient* content of that feed. This gives the amount of the nutrient that a specific feed contributes to the ration. The amount of that nutrient that each feedstuff contributes to the ration is added together. This sum is the total amount of that nutrient in the ration. The cost of the ration may be determined by multiplying the cost of each feedstuff by the amount of that feed fed and adding these amounts together. The calculations for this example are given in Table 3–5.

As shown in Table 3–5, the ration is adequate for *protein* and calcium, but deficient in phosphorus as compared to the animal's requirements. The additional phosphorus needed may be provided by adding monosodium phosphate. Monosodium phosphate contains 22% phosphorus (Appendix Table 5). The amount of monosodium phosphate needed can be determined by using the method known as

TABLE 3–5
CALCULATING THE AMOUNT OF NUTRIENTS IN A RATION AND THE COST OF THE RATION

Feed	Pounds eaten daily	Fraction* of the ration	Protein (%) in feed†	Protein (%) in ration‡	Calcium (%) in feed†	Calcium (%) in ration‡	Phosphorus (%) in feed†	Phosphorus (%) in ration‡	Cost of feed (¢/lb)	Cost of ration (¢/day)
Oats	6	0.40	12	4.8	0.10	0.04	0.30	0.12	6.00	36.0
Alfalfa	9	0.60	17	10.2	1.00	0.60	0.22	0.13	3.50	31.5
TOTAL	15	1.00		15.0		0.64		0.25		67.5
Requirements for weanling (from Appendix Table 2)				14.5		0.65		0.45		

*Fraction of the ration = the amount of each feed eaten divided by the total amount eaten, in this example 6 ÷ 15 = 0.40 and 9 ÷ 15 = 0.60.
†The amount of each nutrient in that feed may be obtained from Appendix Table 1, or more accurately, by analysis.
‡The amount of each nutrient that is contributed to the total ration by any feed equals the fraction of that feed in the ration times the amount of nutrient in that feed, e.g., the fraction of oats in the ration is 0.40, and oats contains 12% protein; therefore, oats contributes (0.40) × (12%) or 4.8% protein to the ration.

the Pearson square (see Example 1 of this section for the explanation of this procedure).

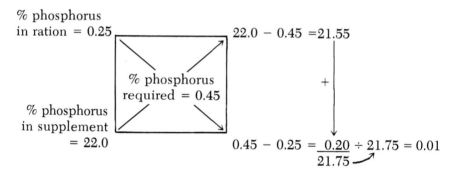

% phosphorus in ration = 0.25

% phosphorus required = 0.45

% phosphorus in supplement = 22.0

$22.0 - 0.45 = 21.55$

$+$

$0.45 - 0.25 = \underline{0.20} \div 21.75 = 0.01$
21.75

As determined using the Pearson square, to increase the phosphorus content of the *ration* to 0.45%, monosodium phosphate is needed in an amount equal to 0.01 of the total feed intake of 15 lbs/day, that is (0.01) × (15 lbs/day) = 0.15 lbs/day, or (0.15 lbs) × (16 oz/lb) = 2.5 oz/day.

The amount of monosodium phosphate needed could be added to the *grain* each day. If the grain does not contain molasses, it may be necessary to dampen it with water to prevent the monosodium phosphate from settling out. Instead of adding monosodium phosphate to the grain daily, it could be added at the feed mill when the *concentrate* is mixed. The amount of monosodium phosphate to be added is the amount needed in the total *ration* (0.01) divided by the amount of grain in the total ration (40%), or 0.025. This would be (0.025) × (2000 lbs/ton), or 50 lbs of monosodium phosphate/ton of grain mix. This same procedure can be used to analyze the nutritive content and to balance any ration.

Example 3

A 600-lb weanling on good quality timothy grass pasture is being fed 4 pounds of a commercial *sweet feed* each morning and evening. The tag on the sweet feed states that it contains greater than 12% crude *protein*. *Cereal grains*, and grain products, but no *mineral supplements*, are included in the list of ingredients. Upon examination, it is found to contain corn, other cereal grains, and molasses. A good estimate of the nutritive content of the feeds as given in Appendix Table 1 would be: 8% protein, 0.25% calcium, and 0.20% phosphorus in the pasture grass; and 12% protein, 0.05% calcium, and 0.25% phosphorus in the sweet feed. If more accurate values were needed, the feeds would have to be analyzed.

The weanling will eat an amount of feed equal to 3.0% of its body weight per day (Appendix Table 2); in this example, (0.03) × (600 lbs) = 18 lbs of feed per day. Eight lbs per day of sweet feed is being fed, which means 10 lbs of pasture grass per day is being eaten. Based on the amount of each feed eaten and the nutritive content of each, the amount of *nutrients* in the total *ration* can be determined as shown in Table 3–6.

TABLE 3–6
CALCULATING THE AMOUNT OF NUTRIENTS PRESENT IN THE TOTAL RATION

	lbs/ day	Fraction* of the ration	Protein (%) in feed†	Protein (%) in ration‡	Calcium (%) in feed†	Calcium (%) in ration‡	Phosphorus (%) in feed†	Phosphorus (%) in ration‡
Pasture	10	0.556	8	4.44	0.25	0.14	0.20	0.11
Sweet feed	8	0.444	12	5.33	0.05	0.02	0.25	0.11
TOTAL	18	1.000		9.7		0.16		0.22
Requirements for weanling (from Appendix Table 2)				14.5		0.65		0.45

* † ‡ See Table 3–5 for explanation of how these values were obtained.

As shown in Table 3–6, the *ration* is deficient in *protein*, calcium, and phosphorus. First, one should correct the protein deficiency by feeding a protein *supplement* such as soybean meal (SBM). This supplement contains 44% protein (Appendix Table 1). The amount needed to increase the protein content of the ration from 9.7% to 14.5% is determined by the Pearson square (see Example 1 of this section for the explanation of this procedure).

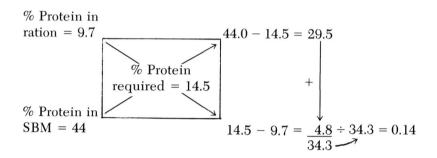

% Protein in
ration = 9.7   44.0 − 14.5 = 29.5

% Protein
required = 14.5   +

% Protein in
SBM = 44   14.5 − 9.7 = 4.8 ÷ 34.3 = 0.14
                              34.3

Therefore, to increase the amount of protein to the desired level, SBM must constitute 0.14 of the *ration*. Since the total amount of feed being consumed per day is 18 lbs, this amount is multiplied by the proportion of SBM needed in the ration (0.14) which means about 3 lbs of SBM should be fed daily. Since SBM is a *concentrate*, substitute this amount for the *sweet feed*. The nutritive content of this *ration* would then be calculated as shown in Table 3–7.

TABLE 3–7
CALCULATING THE AMOUNT OF NUTRIENTS PRESENT IN THE TOTAL RATION

	lbs/ day	Fraction* of the ration	Protein (%) in feed†	Protein (%) in ration‡	Calcium (%) in feed†	Calcium (%) in ration‡	Phosphorus (%) in feed†	Phosphorus (%) in ration‡
Pasture	10	0.555	8	4.4	0.25	0.14	0.20	0.11
Sweet feed	5	0.278	12	3.3	0.05	0.01	0.25	0.07
SBM	3	0.167	44	7.3	0.35	0.04	0.60	0.10
TOTAL	18	1.000		15.0		0.19		0.28
Requirements for weanling (from Appendix Table 2)				14.5		0.65		0.45

*†‡See Table 3–5 for explanation of how these values were obtained.

Correct the calcium and phosphorus deficiency shown in Table 3–7 by feeding a *supplement* such as dicalcium phosphate, which contains 24% Ca and 18% P (Appendix Table 5). The amount needed is calculated by using the Pearson square.

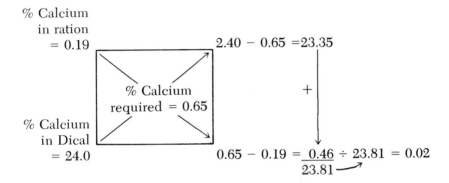

% Calcium
in ration
= 0.19

2.40 − 0.65 = 23.35

% Calcium
required = 0.65

+

% Calcium
in Dical
= 24.0

$0.65 - 0.19 = \dfrac{0.46}{23.81} \div 23.81 = 0.02$

The amount of dical needed would be:
(0.02) × (18 lbs of feed/day) = 0.36 lbs/day or
(0.36 lbs/day) × (16 oz/lb) = 6 oz/day

Check to see if the *ration* now supplies the horse's requirements as shown in Table 3–8.

TABLE 3–8

CALCULATING THE AMOUNT OF NUTRIENTS PRESENT IN THE TOTAL RATION

	Pounds per day	Fraction* of the ration	Protein (%) in feed†	Protein (%) in ration‡	Calcium (%) in feed†	Calcium (%) in ration‡	Phosphorus (%) in feed†	Phosphorus (%) in ration‡
Pasture	10	0.545	8	4.36	0.25	0.14	0.20	0.11
Sweet feed	5	0.272	12	3.27	0.05	0.01	0.25	0.07
SBM	3	0.163	44	7.19	0.25	0.04	0.60	0.10
Dical	0.36	0.020	0	0	24.00	0.48	18.00	0.36
TOTAL	18.36	1.000		14.8		0.67		0.64
Requirements for weanling (from Appendix Table 2)				14.5		0.65		0.45

*†‡See Table 3–5 for explanation of how these values were obtained.

Check the *calcium : phosphorus ratio* in the diet to ensure that it is in the range of 1:1 to 3:1 by dividing both the amount of calcium (0.67%) and the amount of phosphorus (0.64%) present in the total *ration* by the amount of phosphorus (0.64%).

$$Ca:P = 0.67:0.64 = \frac{0.67}{0.64} : \frac{0.64}{0.64} = 1.05:1$$

Instead of feeding 5 lbs of the *sweet feed*, 3 lbs of SBM, and 6 oz of dical daily, these could be mixed together at the feed mill. Then the content of the *concentrate* mix in a one-ton batch would be as follows:

	lbs/day				% of Concentrate mix	lbs/ton
Sweet feed	5	÷	8.36	=	60	1200
SBM	3	÷	8.36	=	36	720
Dical	0.36	÷	8.36	=	4	80
TOTAL	8.36				100	2000

Chapter **4**
Problems Associated with Feeding

FOUNDER

*Founder** is another name for laminitis, which is an inflammation of the laminae of the horse's foot. Laminae are located between the bone and the hoof, and contain blood vessels that nourish the hoof. When inflamed, the laminae between these two rigid structures swell, causing pressure, pain, and tissue damage. Although total blood supply to the hoof is increased, most of this blood is shunted to the heel. Circulation to the toe is decreased, resulting in ischemic necrosis of the laminae at the front of the hoof.

Founder may be caused by ingestion of excessive amounts of *grain*, lush green *forage*, or, in an overheated horse, large amounts of cold water. It may result from *infectious* conditions, such as *uterine* infection after foaling, or severe pneumonia. It may also occur as a result of concussion to the feet from hard work or running on a hard surface.

Horses differ greatly in their susceptibility to *founder*. A sudden increase in the amount of *grain* ingested, as compared to that the horse is accustomed to, is a common cause. A horse that has been receiving a large amount of grain can tolerate more grain than a horse that has not been receiving any. Horses are more susceptible to concussion-induced founder if they are not accustomed to a hard surface. Once founder occurs, the horse is more susceptible to its recurrence, regardless of cause.

Founder may occur when lush green *forage* is consumed. This occurs most commonly in ponies on lush green pasture containing clover or alfalfa. It may also occur on grass pasture, or when lush green hay is fed. Although it is more common in ponies, it may occur in any breed. Affected animals are usually overweight, and have a heavy crest on the neck caused by excess fat; however, it may also occur in

---

*Words in italics are defined in the glossary.

horses that are at optimum body weight when the *roughage* being fed is changed from poorer quality to excellent quality, such as to leafy alfalfa hay or pasture.

The signs and effects of *founder* are the same regardless of the cause. The front feet are most severely affected. The back feet may or may not be involved. The horse's feet, particularly the toes, are sore and painful. When all four feet are affected, the horse tends to lie down for extended periods. When standing, the horse attempts to keep as much weight as possible off the toes, and stands on the heels. To do this, the affected feet are placed farther forward than normal. The hind feet are placed well under the body, and the fore feet are placed forward with the weight on the heel of the foot (Fig. 4–1). When severely affected, the horse is reluctant to move, and may show evidence of pain, such as anxiety, trembling, increased respiratory rate, and fever (normal respiration rate is 8 to 16 per minute and normal temperature is 99 to 101° F, or 37.2° C to 38.3° C).

Signs of grain *founder* usually do not appear for 12 to 18 hours after ingestion of the *grain*. This often leads to the belief that the horse will not be affected by the excess grain intake, but after this period, laminitis, diarrhea, and signs of severe pain occur. Once these signs occur, it may become difficult or impossible to prevent permanent foot damage and lameness, or in severe cases, to save the horse's life. **When it is known or suspected that a horse has consumed more grain**

Fig. 4–1. Stance typical of the foundered horse or the horse with navicular disease. The feet are placed farther forward than normal to keep as much weight as possible off the toes. The horse shown had navicular disease in the front feet. (Courtesy of Corey Lee Lewis, Fort Collins, Colorado.)

than accustomed, it is an emergency situation requiring immediate veterinary care. **Do not wait to see if signs develop.** You are gambling with the life of the horse or with its permanent lameness. Grass, infection, cold water, and concussion-induced founder are generally less acute than that caused by excessive grain intake, so that the signs of severe pain are lessened. However, identical foot changes may occur (Fig. 4–2).

As a result of *founder*, the hoof wall grows more rapidly than normal because of chronic inflammation, and the feet may develop a long toe that curls up at the end. As the hoof grows out, a series of heavy rings usually develops on the hoof wall because of inflammation of the *coronary band* (Glossary Fig. 1). These rings are usually present for the life of the horse. The hoof may separate from the foot, leaving a crack between the hoof wall and the sole. This is referred to as *"seedy toe."* This crack may allow organisms to enter the foot, causing infections.

The foot changes caused by *founder* may lead to a downward rotation of the third phalanx (most distal bone of the foot). This

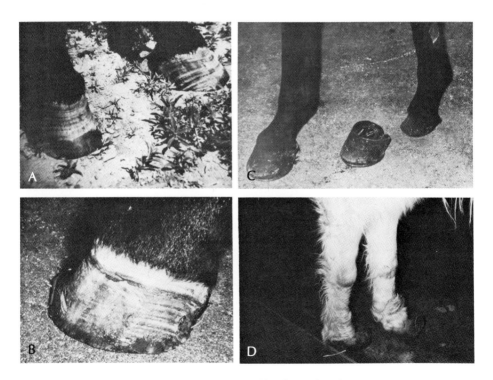

Fig. 4–2(A,B,C,D). Rings on the hoof wall produced by chronic founder (A,B). In severe cases the hoof may separate from the underlying laminae, resulting in the loss of the hoof (C) or the toes may curl upward (D).

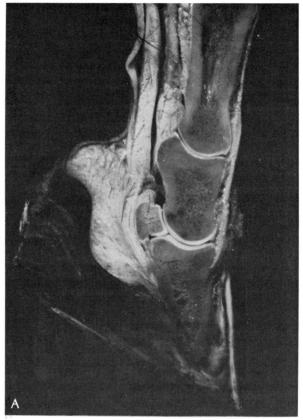

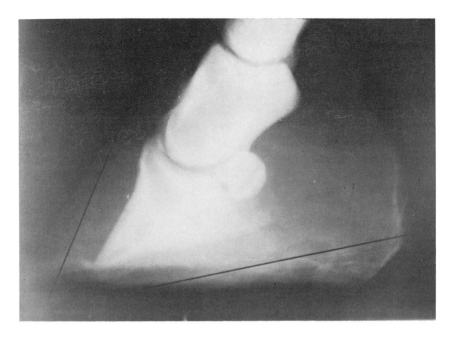

Fig. 4–4. Radiograph showing rotation of the third phalanx (most distal, or bottom, bone of the foot) due to chronic founder. The toe of the third phalanx rotates downward. The black lines are parallel to the surfaces of the third phalanx. In the normal foot these would also be parallel to the sole and the front of the hoof instead of rotated as shown here.

rotation may cause the toe of the third phalanx to push through the sole of the foot (Figs. 4–3, 4–4, 4–5).

To prevent *founder,* eliminate the factors responsible. It is safest not to increase concentrate intake any more rapidly than at a rate of 0.5 lb (0.25 kg) per day. Treatment of founder due to excessive grain intake includes orally giving mineral oil to help expel the grain and antibiotics to kill the bacteria in the intestinal tract, therefore preventing their fermentation of the grain. Giving Butazolidin to decrease

←—

Fig. 4–3(A,B). Rotation of the third phalanx due to chronic founder (A) as compared to the normal (B). The feet have been sawed down the center to demonstrate this. The bottom of the foot is the left bottom surface, the front of the foot is on the right, and the heel, on the left. The digital cushion, which acts as a shock absorber, is the white fatty-appearing tissue at the heel. The bones and joints from the sole up are the third phalanx, coffin joint, second phalanx, pastern joint, first phalanx, and (not shown) the fetlock joint, and cannon bone. The small bone immediately behind the coffin joint is the navicular, or distal sesamoid bone. In the normal foot (B), the distance between the front of the third phalanx and the hoof wall is even, whereas when the third phalanx rotates downward (A), the distance between it and the hoof wall is greater toward the bottom of the foot. This can also be seen on radiographs (Fig. 4–4).

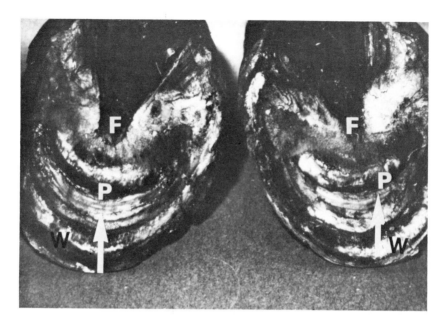

Fig. 4–5.   Rotation of the third phalanx (P) (most distal or bottom bone of the foot) due to chronic founder resulting in its penetration of the sole of the foot. The third phalanx is the black, crescent-shaped structure in the center of the sole approximately one-half way between the apex of the frog (F) and the hoof wall (W). Horses so affected usually must be euthanized for humane reasons.

pain and hypertension and standing the horse in sand to increase pressure on the sole of the foot to assist in preventing downward rotation of the third phalanx may also be beneficial. In grass founder, if the horse or pony is overweight, the weight should be reduced and the animal never again allowed to gain excessive weight. Weight reduction may be accomplished by withholding feed for several days (up to 2 weeks), with water always available and then feeding ½ to 1 lb per 100 lbs of body weight per day (½ to 1 kg/100 kgs/day) of *straw* or a low-energy hay, such as a very mature grass hay, but no grain. Feed in this manner until optimum body weight is attained. Horses or ponies affected by grass founder should never be allowed access to lush, green *forage*, hay, or pasture. Feeding thyroid hormone or iodinated casein has also been recommended.

COUGHING, HEAVES, AND BLEEDERS

Chronic coughing is usually due to *bronchitis*, which may lead to *heaves*. Heaves, or *pulmonary emphysema*, requires additional effort for the horse to exhale, because the alveoli of the lung have ruptured and fibrosed. In a horse with heaves, abdominal muscles must be used

Fig. 4–6(A,B). Heave Line. Note the line formed by the abdominal muscles that runs from the middle of the flank forward and down the rib cage toward the point of the elbow (Glossary Fig. 1). This is called a heave line because it develops in a horse with heaves, or pulmonary emphysema, due to an increase in size of the abdominal muscles used to force air out of the lungs. The bay horse was straining at the time the picture was taken, making the heave line more visible. Although the pinto had more severe heaves, the heave line isn't as visible.

to exhale, which increases their size, resulting in what is referred to as a *"heave line"* (Fig. 4–6A,B). Chronic bronchitis and pulmonary emphysema are also some of the major causes of *"bleeders."*[8] Bleeders are horses that bleed from the nostrils during exercise. This blood is coughed up from broken blood vessels in the lungs. Lung hemorrhage during exercise may also be associated with previous respiratory infection. Many respiratory diseases leave some scarring of the lungs. This portion of the lungs has decreased flexibility and fails to move with adjacent lung tissue, resulting in tearing of tissue and hemorrhage.

Recent surveys suggest that from 40 to 75% of racing Thoroughbreds have some degree of lung hemorrhage after a hard race. Although the incidence is higher with increasing age of the horse and distance of the race, it may occur in 2-year olds in training but never raced, and in distances as short as 3 furlongs. Lung hemorrhage may be a major reason why some horses do not run as well as expected. Only a small percentage of the horses show evidence of blood at the nostrils. Without a direct endoscopic examination of the lungs, it is difficult to determine whether lung hemorrhage occurred during exercise. Affected horses may have a distressed or anxious expression, may occasionally cough, and constantly swallow the blood as it leaves the respiratory tract. The drug furosemide (150 to 250 mg just before a race) is effective in decreasing lung hemorrhage, and will increase the racing performance of affected horses.

Fungal spores, present in some hay or *straw,* may cause a *chronic allergy* resulting in *bronchitis* and *pulmonary emphysema.*[8,15] These microscopic spores may be present even in the finest quality *roughage.* The hay or straw does not have to be visibly musty. These spores are most common in *legumes* such as alfalfa, but may be present in other roughages.

Chronic bronchitis and *pulmonary emphysema* may also be caused or aggravated by inadequate stable ventilation. Ideally, affected horses should be kept on pasture or in open yards for as much as the time as practical. When the horse is stabled, it should be kept in a large, airy box stall with the top door open. All possible ventilation facilities should be fully utilized. Ventilation louvers in the ceiling should be open, and everything should be done to ensure an adequate circulation of air (at least 8 to 10 changes per hour). The use of mechanical ventilation may be necessary. Prior to occupation, the box stall should be given a thorough cleaning, and all dust and cobwebs removed from rafters, windowsills, and other such places. The airspace of the box stall should be isolated from other horses that are not being kept on a fresh-air system; i.e., the partition walls should be complete and air-tight so that the affected horse's atmosphere is not polluted by dusty air from other areas of the stable. This is obviously impractical in some circumstances, and economic and emotional factors must be evaluated. To further decrease dust and *fungal spores*, peat or wood shavings should be used for bedding (do not use black walnut shavings—see discussion of bedding, Chapter 10). All *straw* should be eliminated from the environment. Hay *cubes*, wafers, or pelleted feeds are preferred. If loose hay is used, it should be soaked by putting it in a haynet and immersing it completely in a tank of clean water for a few minutes prior to being fed. As an alternative, hay should be well hosed before being placed inside the stall. If the horse has a nasal discharge, it is best to feed on the ground so that the nasal passages will drain when the head is lowered.

Supplementary medication with *expectorants, chemotherapeutic agents, antibiotics* and *bronchiolytic* agents may be indicated in selected cases. *Nebulization* therapy (using a vaporizer) may also be helpful. However, it should be stressed that the most important items in the management of this common scourge are **fresh air** and a dust-free environment and feeds (pasture, *cubes*, wafers, or pellets) (Fig. 4–7).

Horses with *heaves* will often cough when they are first brought out for exercise, but as long as the coughing does not persist, regular and slow exercise is indicated. With increased fitness, faster exercise is possible, and many horses with *chronic bronchitis* and *emphysema* are capable of doing useful work for many years. Nevertheless, the problem will return if they are exposed to dusty air, so vigilance in "dust-free" stable management must be maintained.

Bleeders are a serious problem, especially in the horse whose work is competitive (Fig. 4–8). In addition to following the recommendations given for the horse with a *chronic* cough or *heaves*, the bleeder must be allowed at least six weeks and preferably two to three months convalescence.[8] During this time, no cantering or galloping should be permitted. In the case of racehorses, it is sometimes necessary to take them out of training for the remainder of the season. If the horse is to be kept in training, walking exercise may be started two weeks after the last hemorrhage, followed by trotting exercise after four weeks. Slow cantering may be started after six weeks, but the horse should not be galloped for two months from the time of the lung hemorrhage. If bleeding recurs, the horse should be taken out of training for the season and turned out to pasture. If bleeding persists the following season, in spite of the recommended management changes having been conscientiously followed, the horse should be regarded as a permanent respiratory invalid, and be retired from racing. These horses are usually capable of carrying out less strenuous work without bleeding.

Chronic bronchitis and *emphysema*, even in mild form, are serious conditions for an animal whose work depends on an ability to perform

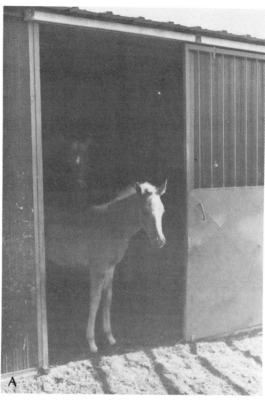

Fig. 4–7(A,B). If the horse must be housed, it is best to have a run, and to leave the box stall door always open to ensure fresh air, decrease dustiness in the stall, and let the horse go in or out of the stall at will. If the stall does not have a run, the top half of the box stall door should be left open to the outside.

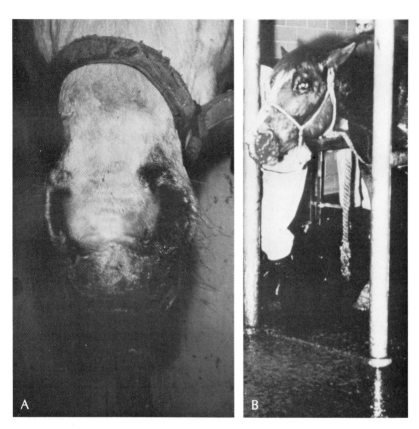

Fig. 4–8(A,B). Bleeding from the nose, or epistaxis. This may be caused by trauma, such as from a tube passed through a nostril into the stomach, blood coagulation disorders, or from blood coughed up from broken blood vessels in the lungs. Horses in which this occurs during physical exertion are called bleeders.

athletically, especially if the work is highly competitive. The lung changes associated with this disease are only partially reversible. Most horses are left with a degree of permanent lung damage. Their ability to do competitive work depends on the severity of the damage. Inevitably, horses with a substantial degree of chronic lung disease cannot be expected to perform as high class athletes, and their retirement from competitive work, such as racing, may be necessary on humane as well as on economic grounds. For these reasons, as in all diseases, an ounce of prevention is worth many pounds of treatment.

WOOD CHEWING AND CRIBBING

Wood chewing is a bad habit that some horses acquire. It usually causes little harm to the horse, although splinters can cause *buccal* infections and excessive tooth wear. Some horses swallow the wood, although others do not. It is an undesirable habit primarily because of the damage it causes to fences, feeders, and barns (Fig. 4–9). This damage can usually be decreased by putting a large tree stump in the paddock for the horse to chew on.

Wood chewing is sometimes incorrectly referred to as *cribbing.* Cribbing is a habit of force-swallowing gulps of air. The cribber usually grasps an object with its incisor teeth, then pulls its neck back into a rigid arch as it swallows air (Fig. 4–10A). It is more harmful to horses than wood chewing since it may cause gastric upset, *colic,* and in some horses, weight loss and poor condition. Fastening a strap several inches (5 to 7 cm) wide snugly around the throatlatch will often prevent the horse from arching its neck, and thus prevent cribbing (Fig. 4–10B,C)

A horse that cribs is sometimes incorrectly referred to as a *wind sucker.* A wind sucker is a mare that aspirates air (and usually also fecal material) into the vagina, particularly when running. This results in an inflammation of the vagina and sometimes of the uterus. It is corrected surgically by suturing the upper portion of the lips of the vulva together. This is called a Caslick's operation.

Fig. 4– 9.   Effects of wood chewing by horses. Although wood chewing is not normally harmful to the horse, it is damaging to fences, barns, and stalls. Creosote and similar substances may be applied to chewed wood, as was done here. This may lessen wood chewing, but the substance must be reapplied at least annually. An electric wire may also be placed just above the top rail. Note the white insulator on the post for this purpose. Preventing boredom in the horse through frequent use, freedom, and companionship are the best means of preventing or lessening wood chewing, as well as other stable vices such as kicking the stall and cribbing.

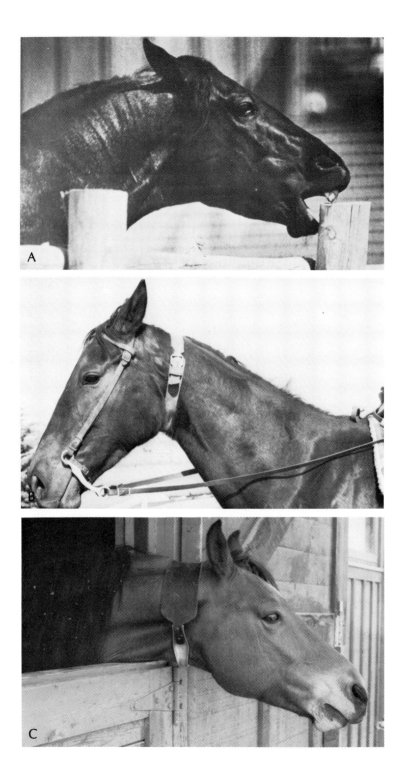

Fig. 4–10(A,B,C). Cribbing. A vice in which the horse places its upper incisor teeth on a solid object, pulls down, arches its neck, and swallows gulps of air. To prevent this, a wide leather strap, called a cribbing strap, is fastened snugly around the throatlatch. It may or may not contain a metal piece at the throat.

TABLE 4–1
EFFECT OF HAY VERSUS PELLETING AND FEEDING FREQUENCY ON WOOD CHEWING[32]

Feeding	Disappearance of pine board in grams/day/horse
Hay	81
Pellets	312
Once/day	211
6 times/day	103

It is important to prevent *cribbing* and *wood chewing* as quickly as possible because they are bad habits frequently copied by other horses. Wood chewing may increase greatly (as shown in Table 4–1) when pelleted or *cubed* feeds are used as the only *roughage*.[32] Increasing the firmness of the pellet may decrease wood chewing. The firmness of the pellet can be adjusted by altering its molasses content. Too much molasses makes a soft pellet, and too little results in a pellet that crumbles easily (7.5 to 10% is generally about the right amount). If wood chewing occurs when pellets, cubes, or wafers are fed, at least ½ lb of long stem hay per 100 lbs of body weight per day (½ kg/100 kg) should also be fed. Increasing the frequency of feeding will also reduce wood chewing (Table 4–1).

Horses chew wood out of boredom, from habit, or because they like its taste, usually not because of any nutritional deficiency other than a lack of adequate *fiber* or *roughage*; but even with adequate fiber, some *wood chewing* may occur. Although deficiencies of a number of nutrients (such as phosphorus, salt, and copper) may result in an abnormal appetite and an increase in wood chewing and consumption of dirt, in the vast majority of cases of wood chewing no nutritional deficiency is present. However, ensure that trace-mineralized salt is available and that the horses' phosphorus requirements are met as described in Chapter 3. If there is any doubt as to whether the ration is adequate in phosphorus, a salt-mineral mix containing calcium and phosphorus should be made available (Appendix Table 5). Although a potassium chloride deficiency is not responsible for wood chewing, adding 1 to 3 oz of potassium chloride to the ration daily or mixing 25 to 50% potassium chloride with the salt-mineral mix that is available to the horses may decrease wood chewing in some, but not in most cases. Potassium chloride can be purchased from some feed stores, laboratory supply outlets, and most grocery stores, where it is usually called salt substitute. However, the best means to prevent or reduce stable vices, such as wood chewing, *cribbing*, and kicking the stall, is to eliminate boredom. The best means of doing this is to decrease confinement, provide plenty of exercise, and provide companionship (from you or from other animals). Applying creosote or other compounds to chewed boards may help decrease wood chewing, but must be repeated often, at least annually.

FESCUE TOXICITY

Fescue is a common pasture grass and a good *forage* for horses. Its major advantage is that it withstands heavy grazing. Its major drawback is that it may occasionally cause fescue toxicity in horses, cattle, and sheep. This occurs most often in warm, humid climates, and usually when horses are on fall pasture regrowth, particularly when autumn rains follow a dry summer. In contrast to previous beliefs, controlled studies have demonstrated that nitrogen fertilization of fescue pasture does not increase the occurrence of fescue toxicity.[97] Symptoms most frequently occur when little or no feeds other than fescue are eaten. It usually happens in the late fall and winter, with the onset of snow and cold, or soon thereafter, or during rainy, overcast weather. Symptoms include (1) rough, dull hair coat, (2) *emaciation*, (3) listlessness, (4) arched back, (5) elevated rectal temperature (up to 106° F, or 41° C), (6) increased respiration and heart rates (normal in the mature horse is 8 to 16, and 28 to 40 per minute, respectively), (7) diarrhea, and (8) cloudy *corneas*. Abortion late in pregnancy, or the birth of weak foals that may die have also been attributed to fescue toxicity. The most common effects in the horse, however, in the order of their occurrence, are a decrease or absence of milk production, prolonged gestation, abortion, and a thickened placenta.[98] Symptoms similar to *ergotism* may occur in cattle and sheep, and rarely in horses. These symptoms are produced by a compound that causes blood vessel constriction and decreased blood flow to the extremities, resulting in *gangrene* of the limbs, tail, and ears. When toxicity in cattle and sheep occurs in the summer instead of the fall, there tends to be more diarrhea, skin lesions, loss of hair and increased efforts to dissipate body heat instead of sloughing of extremities.

In areas where this is a problem, pregnant mares should be removed from fescue pasture during at least the last two to three months of pregnancy. When animals remain on fescue pastures, supplemental use of other feeds will reduce the incidence and/or severity of fescue toxicity. *Thiabendazole* at doses of 5 g/100 lbs of body weight (0.11 g/kg) orally at 7-day intervals has been shown to have a preventative effect in cattle grazing toxic fescue pastures. Thiabendazole is an antifungal agent, so fescue toxicity may be caused by a fungus. It is thought at present that Fusarium tricinctum, or a mycotoxin produced by it, is the fungus involved. Selenium supplementation during pregnancy to mares grazing fescue pasture appears to be beneficial in decreasing fescue toxicity.[97]

NITRATE TOXICITY

The nitrate levels for feed and water given in Table 4–2 are considered safe for *ruminants*. Horses are less susceptible to nitrate toxicity than are ruminants, but for safety's sake it is recommended that horses' feed and water not contain greater than these amounts. *Roughages* containing less than 1% (10,000 ppm) nitrate in their *dry matter* content are generally considered safe if the animal is receiving a good *ration* and is in good condition nutritionally. It is recommended, however, that the total ration for either horses or ruminants contain a maximum of one half this amount. Roughages containing greater than 0.5% (5000 ppm) nitrate (NO_3) should be fed with a sufficient quantity of *cereal grain* so the total ration contains less than this amount (Table 4–2). Nitrate content of feeds is routinely determined in most feed analysis laboratories. The amount present is reported in a number of ways, and can be converted to nitrate content as shown in Appendix Table 8.

Intake of excessive amounts of nitrates may result in *acute* toxicity. Ingested nitrates are converted to nitrites. Nitrites are absorbed, and convert blood *hemoglobin* to *methemoglobin*. This compound prevents the blood from picking up oxygen from the lungs, which gives

TABLE 4–2
METHODS OF CALCULATING RATIONS FOR FEEDING ROUGHAGES HIGH IN NITRATES SO THAT THE TOTAL RATION CONTAINS 0.5% OR LESS NITRATE

Roughage		Grain
Nitrate (%)	Maximum % of total ration eaten	Minimum % of total ration eaten*
0.5	100	0
0.75	0.5 ÷ 0.75 = 67	100 − 67 = 33
1.0	0.5 ÷ 1.0 = 50	100 − 50 = 50
1.25	0.5 ÷ 1.25 = 40	100 − 40 = 60
1.5	Don't feed, since it would require feeding too much grain.	

*The total amount of feed the horse eats daily is shown in Appendix Table 2. Example: A 1000 lb (454 kg) mature horse at rest will eat 15 lbs (6.8 kgs) of feed daily (1.5% × 1000 lbs). If the roughage contains 0.75% nitrate, 5 lbs (2.3 kgs) of grain daily should be fed (33% × 15 lbs, or 33% × 6.8 kgs).

the blood a chocolate-brown color. To observe this, put a spot of blood on white paper and watch for one to two minutes; the blood may not be chocolate-brown in color immediately upon taking it from the animal. This condition is referred to as acute nitrate toxicity. Clinical signs usually occur within one-half to four hours after excessive nitrate ingestion. These signs include abdominal pain *(colic)*, diarrhea, frequent urination, and signs of hypoxia (inadequate oxygen). Hypoxia causes a rapid, weak pulse, increased depth and rate of respiration, labored breathing, incoordination, muscle tremors, weakness, and a dark bluish tinge to body tissues (cyanosis), noted particularly on the mucous membranes of the mouth. Nitrate toxicity as the cause of death or of abortion may be confirmed by freezing the fluid in the eyeball (aqueous humor) from the dead animal or aborted fetus, and sending it to a veterinary diagnostic laboratory for nitrate analysis. Analysis of stomach or intestinal contents is of little benefit. Nitrate toxicity or high nitrate levels in feed, water, serum, or aqueous humor may also be confirmed using the following field test.

1. Prepare a stock solution.

Add 0.5 g of diphenylamine to 20 ml of water. Then add enough concentrated sulfuric acid to bring the volume to 100 ml. Store in a brown bottle.

2. Test plant material.

Split the stem or stalk to be tested, and place 1 to 2 drops of the stock solution on the split surface. The immediate appearance of a dark blue color indicates excessive nitrate levels. Occasionally, this may occur when excessive levels are not present, and sometimes a black or red color may appear. If a dark blue color occurs, don't use that feed, or have it analyzed to determine the amount of nitrate present, so the amount of it that may be fed without causing harm can be determined.

3. Test water, *serum,* or the aqueous humor (fluid in eyeball).

Dilute one part of the stock solution with 10 parts of water. Add 1 drop of serum, aqueous humor, or the water being tested to 4 drops of the diluted stock solution. The immediate appearance of a dark blue color indicates the presence of nitrate or nitrite at levels as low as 40 ppm, and in the serum or the aqueous humor confirms the presence of nitrate toxicity.

Acute nitrate toxicity is the only confirmed effect of excessive nitrate intake. Abortion, reproductive problems, decreased growth rate, and increased *vitamin A* requirements in horses, cattle, and sheep have frequently been attributed to *chronic* nitrate toxicity. However, the results of numerous studies have demonstrated that the presence of nitrates in the *ration* at levels lower than those necessary to produce acute toxicity, even when ingested for prolonged periods of time, does not cause any of these problems.

Acute nitrate toxicity is treated by methylene blue given intravenously (5 mg/lb of body weight, or 10 mg/kg) at several-hour intervals. Methylene blue assists in converting *methemoglobin* to *hemglobin.*

Nitrate is taken up from the soil by plants, and is utilized to form *protein.* Little nitrate is found in *grain,* but under certain conditions appreciable levels are found in plant stems and stalks. Several factors affect the level of nitrate in plants. Increasing levels of fertilizers containing nitrogen or potassium will increase the nitrate content of most *forage* crops. This does not mean that one should not fertilize. Fertilization may greatly increase forage production. This benefit would more than offset the increase in the forage nitrate content that may be caused by fertilization if other conditions are favorable. Forage nitrate content is also increased by any factor that decreases the rate of plant growth prior to maturity. This includes drought, hot weather, early frost, hail damage, or long periods of cloudy weather. After these periods, a sudden change in growing conditions conducive to plant growth results in plants taking up nitrate while converting little of it to protein. These plants, as well as the *stubble* remaining after harvesting forage or cereal grains, may contain high levels of nitrate. This occurs in stubble particularly if it begins to grow and its growth is stopped by a sudden decrease in temperature.

Under adverse weather conditions, the stalks (not the *grain,* and usually not the leaves) of corn, oats, wheat, barley, rye, sudan, and fescue can contain high levels of nitrate. Weeds and some wild grasses that are capable of nitrate accumulation include careless weed, pigweed, lambsquarter, sunflower, bindweed, thistle, redroot, mintweed, and many others. Most frequently *grazed* vegetables capable of accumulating large amounts of nitrate include sugar beet and turnip tops, lettuce, cabbage, potatoes, and carrots. Legumes probably never accumulate enough nitrate to be toxic.

Excessive levels of nitrate in feeds can be prevented by (1) cutting at later stages of maturity, (2) *ensiling,* which may decrease nitrate content 40 to 60%, (3) not cutting for the first few days after a rain, and (4) cutting on the afternoon or evening of a sunny day. If feeds contain a high level of nitrate, (1) add them to the *ration* in small quantities over several days, (2) feed them with the amount of *grain* given in Table 4–2, (3) feed frequently, and (4) don't feed the feedstuffs damp.

Nitrates may also be present in drinking water. The most common source of nitrates in water is from fecally-polluted surface of subsurface water. High nitrate levels are therefore more common in shallow-well water. Although nitrate toxicity is unlikely when using water containing less than 400 ppm nitrate (assuming that a good *ration* low in nitrates is fed), water should not be used if it contains more than 200 ppm nitrate.

Chapter 5

Feeding and Care for Maintenance or Work

The horse needs 1.5 to 1.75 lbs of average to good quality *roughage**per 100 lbs of body weight (1.5 to 1.75 kg/100 kgs) daily for maintenance (Appendix Tables 1 and 2). Most roughages contain adequate quantities of all *nutrients* to meet the horse's requirements for maintenance. However, some roughages may be deficient in phosphorus, and some grass hays deficient in *protein* (Table 5–1).

When a *protein* deficiency exists, it can be corrected by decreasing the amount of hay and adding *grain*, or a protein *supplement*, to the *ration*. A phosphorus deficiency may be corrected by feeding a salt-*mineral* mix containing phosphorus as the only available salt. Although salt intake may be erratic, this is usually adequate for the mature horse for maintenance or work. If grass *forage* is being consumed, a salt-mineral mix containing roughly equal amounts of calcium and phosphorus is preferred, e.g., ½ salt + ½ calcium phosphate, or any number of commercially available mixes containing 8 to 12% calcium, 8 to 12% phosphorus, and salt (Appendix Table 5). If a *legume roughage*, such as alfalfa, is being fed, this same type of salt-mineral mix may be used; however, a salt-mineral mix higher in phosphorus than calcium is preferred, because legumes contain three to eight times more calcium than phosphorus. A high phosphorus salt-mineral mix may be prepared by mixing equal parts of salt, calcium phosphate, and sodium phosphate. Commercially available mixes containing 6 to 10% calcium and 14 to 18% phosphorus are also available (Appendix Table 5). Some commercial mineral supplements, despite their names, do not contain substantially more phosphorus than calcium. Therefore, **read the label to determine the mineral content of any supplement.** Additional calcium is not needed nutritionally when legume roughages are fed. However, both calcium and salt must be added to the phosphorus-containing minerals to increase their *palatability*, and to prevent them from caking and

---

*Words in italics are defined in the glossary.

TABLE 5-1
NUTRIENTS NEEDED BY THE MATURE HORSE FOR MAINTENANCE OR
WORK AS COMPARED WITH THOSE IN FEEDS[50]

	Protein	Calcium	Phosphorus
	————% in total air dry ration————		
Required:	8.0	0.30	0.20
*Composition of:**	————% in air dry feed————		
Alfalfa (or other legumes)	13-19	0.80-2.00	0.10-0.30
Grass	5-13	0.30-0.50	0.10-0.30
Cereal Grains	9-12	0.02-0.10	0.25-0.35

*For more exact values see Appendix Table 1 for the specific type of grain or roughage being fed; for the most accurate values have the feed analyzed (see Chapter 2).

becoming so hard that consumption is reduced. Since calcium is inexpensive, it adds little to the cost of the mineral mix, and the additional calcium is not nutritionally detrimental. Palatability can also be increased and caking prevented by adding 5 to 10% soybean, cottonseed, or linseed meal to the salt-mineral mix.

Energy requirements increase with increasing physical activity, but the need for additional *nutrients* in the *ration* does not increase with physical activity. Therefore, feeding nutritional *supplements* is unnecessary, since additional *roughage* or *grain* will meet the horse's needs. For extensive physical exertion, the horse may not be capable of eating a sufficient amount of roughage to meet its energy needs. Thus grain, which contains a higher energy density, is needed. A guideline for the amount of grain needed for physical activity is given in Table 5–2. Adjust the amount fed as necessary to maintain optimum

TABLE 5-2
GRAIN NEEDED FOR PHYSICAL ACTIVITY*

Physical activity	Grain per hr. of activity	
	(lbs)	(kg)
Light (e.g., pleasure ride)	0.5-1.5	0.2-0.7
Moderate (e.g., ranch work, roping, cutting, barrel racing, jumping)	2-3	1-1.5
Heavy (e.g., race training, polo)	4 or more	2 or more

*In addition, feed 1.5 to 1.75 lbs/100 lbs body weight (1.5 to 1.75 kg/100 kgs) daily of average or better quality roughage. Adjust the amount fed as necessary to maintain optimum body weight and condition.

body weight and condition. In optimum body weight and condition, the ribs cannot be seen but can be felt, with little or not fat between the skin and ribs.

The same feeding programs given above are applicable for the breeding stallion. Breeding does not require an increase in any *nutrient* except those needed for *energy.* The increased energy needs for the act of breeding itself are small. However, the increased physical activity that may be associated with breeding, such as pacing the run or stall, will increase energy needs. There is no confirmed evidence to indicate that dietary supplementation of **any** substance is beneficial for reproductive performance of either mares or stallions.[50,73] Although vitamin E is frequently recommended, there is no convincing data supporting the belief that it is beneficial for reproductive performance in either mares or stallions, and some data refuting it.[73]

FEEDING DURING COLD OR HOT WEATHER

Even in cold weather horses frequently prefer to be outdoors. Closed stables that are kept warm are generally poorly ventilated. Pneumonia occurs more frequently when horses are either kept in warm stables continually, or are moved from warm stables into the cold. Horses acclimate to cold temperature without much difficulty if they are given the opportunity. However, during the winter the horse

Fig. 5–1. A shed open on one side, as shown in the background, is an ideal shelter for all seasons. It provides shelter from storms, shade during the summer, and good ventilation, which is beneficial in preventing respiratory problems that may occur when hay is fed and stalls with poorer ventilation are used. Pasture providing ample quantities of good forage, and large enough for running and playing is ideal for the growing horse. Exercise is important for good bone and muscle development.

should be provided shelter from wind, sleet, and storms, and during the summer it should have shade from the hot sun available. A shed open on one side works well for all seasons (Fig. 5–1).

A long hair coat is an excellent insulator and provides the first line of defense against the cold. Its insulating value is lost, however, if it gets wet, which is why it is important to keep horses dry in cold weather. Horses kept in warm stables will not grow their winter coat, and those that have, will shed it. Occasionally, a horse raised in a warm climate may require an extended period of time to grow its winter coat the first time it experiences a cold climate. The horse will shed its long winter coat as the temperature becomes warmer in the spring. Adding 2 to 4 oz (57 to 113 ml) of cooking *oil* to the horse's *ration* daily, and grooming frequently will hasten shedding and give the horse a glossy hair coat.

The second line of defense against the cold is *fat;* a layer of fat under the skin is excellent insulation against cold. The horse should be in good condition nutritionally when cold weather begins, and obesity should be prevented during hot weather. Nothing is more miserable than an obese animal without shade during hot weather.

During cold weather the best feeding program is to provide adequate amounts of *roughage* for the horse to eat at will. A greater percentage of the *energy* in roughage is given off as heat than that in *concentrates*. Concentrates are digested and absorbed primarily in the small intestine (see Glossary Fig. 2). Little heat is produced in this process. In contrast, roughages are digested by bacterial fermentation in the *cecum* and large *colon*, and a great deal of heat is produced in this fermentation process. In hot weather this is obviously a disadvantage, whereas in cold weather it is beneficial.

During cold weather, if the horse is allowed to consume all the good quality *roughage* it wants, additional *energy* from *cereal grains* or other *concentrates* is not usually needed. However, if the horse is not able to maintain good nutritional condition, *grain* should be fed. Any cereal grain may be fed, but corn is preferred because of its high energy density. Corn contains twice as much energy as an equal volume of oats. Therefore, a smaller volume of corn is required to provide the additional energy needed. This allows more room in the digestive tract, so that a greater amount of roughage can be consumed.

Contrary to popular belief, corn is not a *"heating feed."* The amount of *energy* from corn which is given off as heat in its digestion, absorption, and utilization is one-third less than that produced from oats. The lower amount of heat produced from corn is more than offset by the increased amount of heat produced from the additional *roughage* which can be consumed when corn is fed. Corn is therefore a good cereal grain to feed during the winter if the additional energy it provides is needed. During hot weather corn is also a good *cereal grain* to feed because it provides the horse's energy needs with minimal heat produced in its utilization, while decreasing the amount of roughage needed to meet the horse's energy requirements. However, corn, or any *concentrate*, should not make up over one half of the *ration* by weight. In addition, any change in the horse's feeding program, either in the type or amount of feed, must be instituted gradually (see When and How to Feed, Chapter 10).

Hot bran mashes are frequently fed during the winter; however, they are of no benefit in keeping the horse warm. Their only benefit is that they increase the consumption of water. Water for horses during cold weather is too often overlooked. The water may freeze over, making it impossible for the horse to get to; and even if it doesn't freeze, when water is very cold, the horse's consumption of it decreases. A decrease in water consumption results in a decrease in feed consumption. Consequently, the horse does not have the *energy* necessary to maintain body temperature and weight during cold weather. Inadequate water consumption may also result in feed becoming impacted in the intestinal tract, which results in *colic.* Although feeding hot bran mashes helps prevent these problems by increasing water consumption, installing heated waterers, and always allowing the horse access to all the water it wants are much more practical means of increasing water consumption and preventing inadequate feed intake and impactions. Ideally, the water temperature should be maintained at 45 to 65° F (7 to 18° C). If an electric water heater is used, the water should be touched daily to make certain the heater is not shorting out in the water. This usually will not create enough of an electrical shock to hurt the horse, but it will prevent water consumption.

FEEDING FOR STRENUOUS PHYSICAL ACTIVITY

For frequent or prolonged strenuous physical activity, such as training to increase physical conditioning, or endurance racing requiring prolonged submaximal physical exertion, the *nutrients*[1] of specific concern in their order of importance, are: (1) water, (2) body salts, or *electrolytes*, and (3) those needed for energy. As discussed at the beginning of this chapter, other nutritional requirements, such as protein, do not increase with physical activity.

The most important requirement for optimal physical exertion is maintaining hydration (water balance). A horse can lose essentially all of the fat and up to one half of the protein in his body, whereas a loss of only 12 to 15% of the body water is fatal. Much lower amounts of water loss result in fatigue and decreased performance. To prevent this, allow the horse to drink as frequently as possible, preferably small amounts at frequent intervals. During physical activity, the horse should be allowed, and encouraged, to drink all the water it wants. When physical activity is stopped, cool the horse by walking. Then allow the horse to *graze*, or eat hay, and to rest for 60 to 90 minutes before watering. After watering, *grain* may be fed. Excessive cold water intake by a hot, nonactive horse may cause *colic* and *founder*. During an endurance race with over two hours between watering places, (or less time if temperature and humidity are high), carry 1 to 2 gal of water (4 to 8 L), and a hat or collapsible pail, so the horse may be watered between scheduled watering places.

During physical exertion, sodium, potassium, chloride, and calcium are lost in the sweat and urine. Loss of the first three *electrolytes* causes fatigue and muscle weakness, and decreases the thirst response to dehydration, so the horse may have little inclination to drink or eat. Dehydration and deficits of these electrolytes are routinely present in the horse exhausted as a result of prolonged physical exertion. It has been noted that in the Tevis Cup Ride, a very difficult 100-mile endurance race under very warm environmental conditions, the horses that finish the ride are those that drink water during the race, while those unable to finish the race stop drinking or drink little during the ride. To prevent this, the electrolytes lost should be replaced and water should be offered frequently.

Excessive calcium loss results in muscle twitching, spasms, and *tetany*. This is referred to as stress tetany, and may be confused with the *tying-up syndrome*, which occurs as a result of muscle *energy* depletion. It may also be confused with *exertion myopathy (azoturia)*, since the clinical signs are similar. However, exertion myopathy occurs within the first 10 to 15 minutes after beginning exercise, whereas stress tetany and the tying-up syndrome usually occur after several hours of physical activity. The conditions occurring in the horse as a result of physical exertion, and the procedures necessary for preventing them are given in Table 5–3.

Thumps, or *synchronous diaphragmatic flutter (SDF)*, is another condition that may occur owing to *electrolyte* losses as a result of physical exertion. A decrease in the *plasma* calcium, chloride, and/or potassium concentrations is thought to increase the irritability of the phrenic nerves. As a result, the electrical activity occurring when the heart beats stimulates these nerves where they pass over the heart. The *diaphragm*, which is stimulated by these nerves, then contracts with each beat of the heart. This is usually observed as sudden bilateral, and occasionally unilateral, movements of the horse's flanks, and sometimes a hind leg, every time the heart beats. This may occur during or following physical exertion, severe diarrhea or *colic*, or following prolonged surgery and anesthesia. It has also been seen in horses which have been eating hay containing blister beetles

TABLE 5–3
DISEASES OCCURRING IN THE HORSE AS A RESULT OF PHYSICAL EXERTION

Disease	Cause	Clinical Signs	Time of occurrence with exercise	Prevention
Exertion Myopathy	Excess lactic acid in muscle		first few minutes	decrease grain, do not confine, warm up slowly, **do not move horse if it occurs!**
Stress Tetany	Excess calcium loss from body	Muscle spasms and reluctance to move		give electrolytes containing calcium during physical activity
Tying-up	Muscle energy depletion		during prolonged physical activity	train to increase conditioning, decrease exertion, rest more often, give water and electrolytes during and after physical activity
Exhaustion	Energy, water and electrolyte deficits	flaccid muscles, decreased food and water consumption, dilated anus, increased temperature		
Synchronous Diaphragmatic Flutter (SDF) or Thumps	Excess calcium, potassium and chloride losses	diaphragmatic contractions and movement of flanks or hind leg in synchrony with the heart beat	during or after physical activity	give electrolytes during and after physical activity
Post Exercise Fatigue	Potassium deficit	fatigue for days		give potassium during and after physical activity

(see Appendix B), and in mares with stress, or lactation, *tetany.* Prevention and treatment is the replacement of lost electrolytes.

The amount of *electrolytes* lost, and, therefore, the benefit of replacing these losses during and after physical activity, will be greater the longer the duration and the greater the severity of the exercise or stress, and the higher the environmental temperature and humidity. In addition, there may be much individual variation between horses in the amount of electrolytes lost under similar situations. Electrolyte replacement is recommended for all horses during prolonged physical exertion, such as endurance racing, even though little benefit may be noted in some horses, especially when there is a shorter duration of physical activity, or lower environmental temperature and humidity. Electrolyte supplementation as described is sometimes necessary and can be tremendously beneficial, and is never detrimental so long as water is available.

Electrolytes lost as a result of physical activity may be replaced by giving the horse 2 oz (57 grams) of 3 parts "lite" salt plus 1 part limestone. This should be given just before an endurance race, at each watering during the race (which, preferably, will not be more than every 2 hours), after the race, and twice on the day following.

"Lite" salt, or low-sodium salt, is one-half sodium chloride, or table salt, and one-half potassium chloride. It can be purchased at most grocery stores. Limestone is calcium carbonate and can be purchased at most feedstores. This is an economical way to provide the *electrolytes* needed. The electrolyte mixture may be added to a few ounces of water or molasses and squirted into the back of the horse's mouth, or it may be added to the feed. Two ounces of the mixture in one gallon (15 g/L) of water provides 50 *mM* of sodium and potassium, 100 mM of chloride, and 20 to 25 mM of calcium, and has an *osmolality* of 200 to 250 *mOsm/L*. Electrolytes should not be added to the drinking water, as doing so can decrease water consumption. Adequate water intake is much more important.

After water and electrolytes, the next *nutrients* needed for physical activity are those that provide *energy*. The three sources of dietary energy are *carbohydrates, fats,* and *proteins.* Several studies have been conducted to determine which of these is the best source of energy for prolonged physical exertion, such as occurs with endurance racing.[27,33,68] Minor differences in some of the physiologic parameters measured, such as blood sugar levels, suggested that high fat diets might be beneficial, and high protein diets detrimental, for endurance activity. However, no difference in physical performance of horses on high fat diets, high carbohydrate diets, or high protein diets was observed.[27,33,68] The physical activity in these studies may not have been extensive enough to show a difference, or no difference may exist.

The following is recommended for feeding the horse for endurance racing and training, and for the horse in training to increase physical conditioning:

(1) Ensure that all the good quality pasture grass the horse wants to eat is available, or provide all the good grass hay, or an alfalfa-grass—hay mix containing 8 to 14% *protein*, that the horse will eat. Some grass hay, even though it may appear to be of excellent quality, may contain less than 7% protein, which is below the horse's protein requirement of 8%, and, therefore, should be fed with a higher-protein-containing feed. Good alfalfa may contain 18% or more protein.

(2) Feed *grain* in amounts necessary to maintain good condition and optimum body weight. The amount of grain fed should not exceed the amount of *roughage* eaten. Although any cereal grain may be fed, I prefer rolled, flaked, or *crimped* corn because it is more consistent in quality, less expensive, because less heat is produced in its utilization, and it is higher in *energy* than other cereal grains.

(3) Feeding bran is not recommended. Bran is a bulky, low *energy* feed. It contains about 25% less energy than corn, and 12% less than oats (Appendix Table 1). Because of its bulkiness and low energy content, it increases the amount of feces and ingesta in the horse's intestinal tract, which is detrimental during physical activity.

(4) As much as one qt (1 L) of vegetable *oil* per day may be added to the horse's *grain*. Research up until now has not proven that this is beneficial; however, if you wish to *supplement* with oil, begin adding the oil to the *ration* at least 3 weeks before endurance racing. Add ½ pt (235 ml) of oil to the ration on the first day. Increase this by ½ pt (235 ml) per day until 1 qt (1 L) is being fed daily. If *palatability* or loose stools are a problem, reduce the amount being fed.

(5) In the last two days before an endurance race, increase the amount of *grain* fed by 1 to 2 lbs (0.5 to 1 kg) per day, and decrease the amount of *roughage* fed by 3 to 4 lbs (1.5 to 2 kgs) per day. On the evening before the ride, feed an additional 1 lb (0.5 kg) of grain. Don't feed the horse either grain or hay on the day of the race, until the race is over. Increasing the amount of grain and decreasing the amount of roughage fed, decreases the amount of ingesta in the horse's intestinal tract and, therefore, the amount of additional weight the horse must carry. The horse's intestinal tract can hold as much as 250 to 300 lbs of ingesta. It takes 50 hours for the undigested *fiber* in roughages to pass through the intestinal tract and be excreted (see Glossary, Fig. 2).

Although extensive nutritional *energy* is needed by the horse during the endurance race, feeding readily available *carbohydrates*, such as those present in *cereal grains*, shortly before or during the race may not be beneficial. In other species, carbohydrates during this period have been shown to be detrimental. They increase the blood sugar concentration which, in turn, causes changes in the secretion of the hormones that decrease the ability to mobilize body *fats, glycogen*, and *protein* for energy utilization. During long term endurance activity, most of the energy needed must be derived from these body stores. A decrease in the animal's ability to utilize them will hasten fatigue and exhaustion.

In short races of only a few minutes duration or less, feeding *grain* to the horse 90 to 120 minutes before the race may be beneficial since it provides a readily available source of *energy*, and mobilization of body energy stores is not necessary, but the benefit of this procedure has not been demonstrated.

(6) Just before an endurance race, and as frequently as practical during the race (preferably every 2 hours or less), offer water, and give 2 oz (57 g) of the *electrolyte* mixture described previously.

(7) Cool the horse as frequently as practical during an endurance race. Allowing the horse to stand in water to just above the knees and hocks, and putting cool water, or towels soaked in water, on the neck, poll, and legs is beneficial. However, care should be taken not to put cold water on the heavy muscles of propulsion, as this may cause muscle spasms and *tetany*. The horse's *energy* expenditure may be increased as much as 10- to 20- fold during an endurance race. This greatly increases the amount of heat produced, which must be eliminated. This is accomplished primarily by the evaporation of sweat from the body surface. Sweating increases water and *electrolyte* losses. As these losses become extensive, sweating is decreased, and the horse's temperature increases. Rectal temperature greater than 106 to 108° F (41 to 42° C) may result in death of body cells, including those in the brain and liver, and be detrimental to the horse. Failure of the rectal temperature to decrease to 102° F (39° C) or less within 10 minutes of rest may indicate that the horse is becoming, or is, dangerously exhausted.

(8) After the race, cool the horse and give it immediate access to all the *roughage* it will consume. After 60 to 90 minutes, let the horse drink all the water it wants, and give 1 to 2 qt (L) of *grain* and two oz (57 g) of the *electrolyte* mixture described previously.

(9) The day after the race, feed in the regular manner, and give 2 oz (57 g) of *electrolytes* that morning and evening.

Assessing Physical Condition

Many parameters have been used to try to assess the horse's physical fitness. The increase in heart rate and respiratory rate with exercise, and the rate of their decrease after exercise, are the most common parameters measured. However, in controlled studies they were found to be of no value.[47] *Cardiac output* and *stroke volume* were also not affected by physical conditioning in the horse.[47] The only parameter found to be of benefit in assessing the physical fitness of the horse is the venous blood *plasma* lactic acid concentration after exercise. Lactic acid is a substance produced by cells (primarily by muscle cells) if there is inadequate oxygen available for those cells. With physical conditioning, there is an increase in blood circulation to the muscle, and an increase in the ability to remove a greater amount of oxygen from the blood. As a result, more oxygen is available to the muscle and, therefore, less lactic acid is produced during exercise.

In one study, in which nonphysically-conditioned horses were trotted 3 to 4 miles (4.8 to 6.4 km), 6 days a week for 3 weeks, then this same distance for 4 days a week, and for 2 days a week ran 1 mile (1.6 km) in 2 minutes 56 seconds to 2 minutes 20 seconds, the blood *plasma* lactic acid concentration 10 minutes after trotting was decreased from an average of 41 mg% on day one to 13.6 mg% on day 92 of training.[47] The blood plasma lactic acid concentrations after exercise declined until day 32 of training. After an additional 60 days of training, there was little further decline in postexercise concentrations. This indicates that by day 32 the horses were in good physical condition.

Blood *hemoglobin* concentration and *hematocrit* are also frequently used to assess the presence of *anemia*, and the horse's general physical condition and nutritional status, particularly with respect to iron and the B-*vitamins*. These parameters are of no benefit in assessing physical condition, and are of little benefit in assessing the animal's nutritional status. Although they are decreased with an iron deficiency, the decrease does not occur until the deficiency is prolonged and severe. Several changes more diagnostic of a deficiency in the *nutrients* necessary for *red blood cell* synthesis (such as iron) occur before there is a decrease in hemoglobin concentration or in hematocrit. These include a decrease in iron-staining in *bone marrow*, and a decrease in *transferrin* iron saturation. The *serum ferritin* concentration has been shown in other species to be the best indication of body iron content, but has not been studied in the horse. If these values cannot be obtained, the hemoglobin concentration may be used. However, the hemoglobin concentration and hematocrit are meaningful **only** if the blood sample is taken immediately after exercise.[54] A large supply of red blood cells is stored in the *spleen*. Complete splenic contraction can increase the hemoglobin concentration and hematocrit by more than 60%;[14] for example, the hemoglobin concentration may be increased from 10 g% to 16 g%, and the hematocrit increased from 30% to 48%. Exercise, apprehension, fear, and excitement, all cause varying degrees of splenic contraction. When a blood sample is taken at rest, the amount of excitement or apprehension and, therefore, the amount of splenic contraction varies, resulting in variable blood hemoglobin concentrations and hematocrits. Only when the splenic reservoir of red blood cells is completely depleted can meaningful hemoglobin concentrations or hematocrits be obtained.[54] To ensure complete splenic contraction, the horse should be exercised until the pulse rate 15 to 45 seconds after the termination of exercise exceeds 100 beats per minute, and the blood sample should be taken within one minute after exercise.[54] When these procedures are followed, the hemoglobin concentration should exceed 15 g%; if it does not, it indicates that the horse is anemic.

Chapter 6
Feeding and Care of the Brood Mare

Three feeding programs are necessary for the brood mare. These are (1) prior to the last three months of pregnancy, while not lactating, (2) during the last three months of pregnancy, and (3) during lactation. Her *energy** needs increase progressively during each of these periods. As a result, the mare will eat more if feed is available and if the *ration* isn't too bulky. The percentage of *protein*, calcium, and phosphorus required in the ration increases, to a greater extent than the amount of energy (Table 6–1). Therefore, simply increasing the amount fed, without changing the ration, may not provide the additional *nutrients* needed.

Failure to provide the pregnant female with adequate nutrition prior to the birth of the young has a profound affect on the offspring of many species. After giving birth, while on a low plane of nutrition, the mare and the cow may have lowered fertility and decreased milk production, which reduces growth rate of the offspring. In cases of severe malnutrition of the dam during pregnancy, more profound, long-lasting effects occur in the offspring, such as skeletal deformities, decreased brain development, diarrhea,[13] increased incidence of disease, and decreased survivability. Offspring of malnourished bitches have decreased vigor at birth and lower survivability.[48] Inadequate protein in the diet during pregnancy has been shown to decrease *cerebral* weight and protein content, as well as to decrease the number of brain cells in the newborn animal.[49] Adequate nutrition of the offspring after birth will not correct these effects of inadequate nutrition prior to birth.[67] Although it has not been proven, it is possible that improper feeding of the pregnant mare may have similar effects on the foal.

The most common errors in feeding the brood mare are overfeeding during pregnancy and underfeeding during lactation. The overfed, obese mare is more likely to have trouble foaling as a result of poor

---

*Words in italics are defined in the glossary.

120

TABLE 6-1
NUTRIENTS NEEDED BY THE BROOD MARE AS COMPARED
WITH THOSE IN FEEDS[50]

	Protein	Calcium	Phosphorus
Requirements:	——% in total air dry ration——		
Last 3 months of pregnancy	10	0.45	0.35
Lactation	12.5	0.45	0.35
After weaning and before last 3 months of pregnancy	8.0	0.30	0.20
Composition of:*	——% in air dry feed——		
Alfalfa (or other legumes)	13–19	0.80–2.00	0.10–0.30
Grass	5–13	0.30–0.50	0.10–0.30
Cereal grains	9–12	0.02–0.10	0.25–0.35

*For more exact values, see Appendix Table 1 for the specific type of grain or roughage being fed, and for the most accurate values, have the feed analyzed.

muscle tone caused by the decreased physical activity often associated with obesity. In addition, she is likely to be losing weight after foaling and during lactation. Mares that are losing weight are more difficult to get bred. In one study, mares that were losing weight took 30 days longer to come into heat than those that were maintaining or gaining weight after January 1.[19] In another study, 8 mares, that during the last 3 months of pregnancy were fed 16 kcal/lb body weight (7.3 kcal/kg) daily averaged 1.3 estrous (heat) cycles per conception as compared to 2.9 estrous cycles per conception in 8 mares fed 20 kcal/lb (9.1 kcal/kg) daily. Thus, **either excess feed intake and obesity or inadequate feed intake during the latter stages of pregnancy greatly decreases the mare's reproductive efficiency.**

If the mare is obese, her weight should be reduced **before** attempting to breed her. During weight loss, she is less likely to come into heat, or to settle if bred. If she is pregnant, weight reduction should be achieved before the last three months of pregnancy. Extensive weight reduction should not be attempted during the last three months of pregnancy because of the effect it may have on the fetus. Weight reduction may be accomplished by not feeding any *concentrates*, and by limiting *roughage* intake. Ideally, mares should be brought into the breeding season at optimal body weight. Increasing the balanced diet two to five weeks prior to the breeding season, so that mares are gaining weight, may improve reproductive performance. Vitamin-mineral supplementation is unnecessary and ineffective in improving the mare's reproductive performance when adequate quantities of average, or better, quality feeds are being fed.[73]

THE NONLACTATING MARE PRIOR TO THE LAST THREE MONTHS OF PREGNANCY

There is little increase in fetal size during the first eight months of pregnancy. During this period, the pregnant, nonlactating mare should be fed in the same way as the mature horse for maintenance or work, as described in Chapter 5. Feeding excessive quantities during this period is a common error.

THE MARE DURING THE LAST THREE MONTHS OF PREGNANCY

Nearly two thirds of fetal growth occurs during the last three months of pregnancy. During this period, the amount of *protein*, calcium, and phosphorus needed in the *ration* increases (Table 6–1).

If good quality alfalfa or other *legumes* are being fed, the additional *nutrient* requirements, with the exception of phosphorus, can be provided by increasing the amount of *forage* available (Table 6–1). The additional amount of feed needed, on a body weight basis, is small. Since the mare's body weight increases about 15% during this period, the amount fed should be increased proportionally. The additional phosphorus which may be needed when a legume *roughage* is fed may be provided by allowing free access to a salt-*mineral* mix containing similar amounts of calcium and phosphorus; however a low-calcium, high-phosphorus salt-mineral mix is preferred (see Chapter 5 and Appendix Table 5). These salt-mineral mixes should be the only available salt. If other salt is available, the mare may not consume adequate quantities of phosphorus.

If grass pasture is available, or grass hay is being fed, generally more calcium, phosphorus, and, occasionally, *protein* are needed (Table 6–1). Green grass pastures usually contain adequate protein to meet the pregnant mare's requirements, whereas dry, mature grass *forage* (pasture or hay), frequently will not. However, if the pasture contains any *legumes*, the forage will generally be adequate in both protein and calcium. A laboratory analysis may be conducted to determine if it is. If the *ration* does contain adequate protein to meet the mare's requirements, the additional phosphorus, and possibly calcium, needed should be provided by allowing free access to a salt-mineral mix that contains roughly equal amounts of calcium and phosphorus (see Chapter 5 and Appendix Table 5) as the only available salt. If *grain* is being fed, 3 oz (85 g) daily of this *mineral* mix may be added to the grain. If the grain does not contain molasses, it may be necessary to dampen the grain when fed to prevent the mineral mix from sifting out. If the feed is dampened, and all of it is not eaten within a few hours, any remaining feed should be removed and discarded. Dampened feed will become moldy rapidly. Moldy feed is unpalatable and may be toxic to the horse.

If the *forage* does not contain adequate *protein* to meet the mare's requirements, a *concentrate* mix similar to those presented in Appendix Table 6 is needed; or one can be formulated as described in Chaper 3 to meet the mare's *nutrient* requirements (Table 6–1). From 0.5 to 0.75 lbs/100 lbs body wt/day (0.5 to 0.75 kgs/100 kgs/day) of the concentrate mix should be fed with the amount of grass hay needed for maintenance.

If the mare is overweight, no additional feed over that needed for maintenance should be fed during this period. However, the *mineral* mixes recommended should be provided.

THE LACTATING MARE

Lactation greatly increases the mare's requirements for *nutrients*. *Energy* needs are increased two-fold, and the requirements for *protein*, calcium, and phosphorus increase (Table 6–1). A deficiency in any one of these nutrients decreases the amount of milk produced but has little or no effect on the concentration of these nutrients in the milk. A normal, healthy 900- to 1200-lb (400 to 550 kg) mare, when properly fed, will produce from 25 to 30 lbs (11 to 14 kg) of milk per day during the first three months of lactation. By five months of lactation, the amount produced will decrease by only about 5 lbs (2.3 kg) per day. It takes about 1200 kilocalories of digestible energy to produce 1 lb (0.45 kg) of milk. The composition of mare's milk is given in Table 8–2. Each mare should be fed as much as necessary to maintain proper body weight and condition. A good indication of proper weight and condition is the amount of tissue covering the ribs. With a summer hair coat or a wet winter hair coat, the ribs should not be seen but should be easily felt without feeling any appreciable amount of fat between the ribs and the skin.

High-energy requirements for lactation, or during the later stages of pregnancy, in the face of a fall in the *nutrient* quality and availability of feed, may result in inappetence, progressive drowsiness, muscle fasciculations, diarrhea, ventral *edema*, and death. This condition occurs most frequently in ponies, and is characterized by a marked increase in blood lipids. Extensive lipidosis and vascular blood clots may be found on postmortem examination. Treatment and prevention is to ensure the availability of adequate quantities of good quality feeds during the later stages of pregnancy and lactation.

During lactation, if good quality alfalfa or other *legumes* are being fed, additional *nutrients* may be provided by increasing the amount, and by allowing free access to a calcium-and-phosphorus-containing

Fig. 6–1. Mares and foals on pasture. Lots of exercise and play are important for good bone and muscle development by the growing horse. A lack of exercise predisposes the foal to contracted flexor tendons. In addition, good green, growing pasture containing ample quantities of forage, along with a salt-mineral mix containing 8 to 16% of both calcium and phosphorus (Appendix Table 5), available for free-choice consumption as the only salt, will meet all of the lactating mare's nutritional needs.

salt-mineral mix (see Chapter 5 and Appendix Table 5) as the only salt available (Fig. 6–1). During lactation, it is difficult to overfeed with *roughages*. Feed as much hay as the mare will eat. This will require at least 2.5 to 3.0 lbs/100 lbs of body weight daily (2.5 to 3.0 kg/100 kg/day).

If grass hay is fed, increasing its amount, and feeding *cereal grain* usually is not sufficient to meet the lactating mare's calcium, phosphorus, and *protein* needs (Table 6–1). A *concentrate* mix similar to

TABLE 6–2
FEEDING PROGRAMS FOR THE LACTATING MARE

Type of roughage fed	Amount of concentrate needed (lbs/100 lbs/day)*	Type of salt-mineral mix preferred†
Alfalfa (or other legumes)	0	High phosphorus, best; balanced calcium and phosphorus, acceptable
Green grass pasture	0	Balanced calcium and phosphorus
Mature grass pasture or grass hay	0.5–1	Trace-mineralized salt‡

*or kg/100 kg body weight/day needed when all of the roughage the mare will eat is available.
†See Chapter 5 and Appendix Table 5.
‡When concentrate mixes similar to those given in Appendix Table 6 are fed.

those given in Appendix Table 6 is needed, or one can be formulated as described in Chapter 3 to meet the mare's *nutrient* requirements (Table 6–1). About 1 lb/100 lbs body weight daily (1 kg/100 kg) of the *grain* mix should be provided, in addition to all the grass hay the mare will eat. As the mare's milk production decreases, feed intake should be decreased to maintain proper body weight. A summary of feeding programs for the lactating mare is given in Table 6–2.

Tetany due to a fall in the *plasma* calcium concentration may occur in the lactating mare as a result of the loss of calcium into the milk. Although this is uncommon, the chances of it occurring are increased if, in addition to lactation, the mare is subjected to prolonged stress of any type such as inclement weather, disease, trauma, strenuous physical exertion, or transport. This condition is treated by injecting calcium-containing solutions into the *vein*. This should be done very cautiously, while listening to the heart, and only by a veterinarian. Giving calcium intravenously too fast, or giving too much, will cause sudden death.

To prevent lactation *tetany* in a mare in which it has occurred previously, feed a low-calcium *ration* during the last two to five weeks before foaling. A low-calcium ration for this short a period of time is not detrimental to either the mare or the foal. Immediately following foaling, switch to a high-calcium ration.

A low-calcium *ration* increases the efficiency of intestinal calcium absorption, so that the high amounts of calcium present in the ration after foaling, and needed for milk production, can be absorbed. The low-calcium ration also stimulates the *parathyroid gland,* so it is able to respond more rapidly and effectively to a fall in the *plasma* calcium concentration, and mobilize more calcium more rapidly from the bone to prevent *tetany.*

A low-calcium *ration* is one consisting of a grass *roughage,* any *cereal grain,* and no *mineral* mixes containing calcium. A high-calcium ration is one consisting of alfalfa, or any *legume,* a *concentrate* mix with a calcium-containing mineral added to it, and free access to a salt-mineral mix containing calcium.

WEANING

Weaning should be done in a manner that will minimize stress and excitement for both the mare and the foal. Some foals become frantic when first separated from their mother, and as a result may injure themselves. When weaning, it is always best to (1) move the mare, leaving the foal in familiar surroundings (2) make the separation abrupt, complete, and final, and (3) leave the foal with another horse that it is accustomed to, or place two foals together. If the foal sees, hears, or smells its mother during this period, weaning is more difficult and prolonged. If they cannot be completely separated in this manner, weaning may be accomplished by putting them in separate paddocks so that the foal cannot nurse.

One of the best ways to minimize stress and excitement, and a resulting injury to the foal at weaning, if you have several mares and their foals, or even one mare and foal with at least one other horse, is to remove one mare from the herd, while leaving her foal with the other horses. If the foal cannot see, hear, or smell its mother, is with other horses that it is accustomed to, and is in familiar surroundings, little difficulty is usually encountered. It is best to remove the most dominant mare from the group first or the mare with the biggest, most independent foal. Wait several days, then remove the next most dominant mare from the group, again leaving her foal with the others. Continue in this manner until all of the foals are weaned.

Several days before and after weaning, don't feed the mare any *grain* and if a *roughage* is being fed, decrease the amount to that needed for maintenance (1.5 to 1.75 lbs/100 lbs body wt/day or 1.5 to 1.75 kg/100 kg/day). This decreases milk production and prevents excessive distension of the mammary gland. The more milk the mare is producing, the more beneficial this is. If the mammary gland or udder becomes distended and painful, rub it with camphorated oil to keep it soft and pliable. Don't milk it out, since this will stimulate more milk production.

Chapter 7
Foaling

The length of pregnancy in the mare is usually 335 to 345 days, although it is not unusual for a mare to go as long as 365 days. Any birth under 325 days is premature, and birth at less than 300 days is not usually compatible with life. Most mares repeat their gestation schedule, so the length of every gestation should be recorded. Any marked variation in this schedule should be investigated.

PREPARATION

Prior to foaling, thoroughly clean the foaling stall or area. Make sure the area is warm, dry, clean, and of adequate size for the mare and foal. The stall should be at least 14 by 14 ft (4.3 m), with walls solid from the floor to at least 4 ft (1.2 m) high, and free from sharp or protruding objects. *Straw** bedding is preferred over wood or peat shavings, or sand, as these are more easily drawn into the vagina during foaling, and are much more abrasive. The signs of approaching *parturition* in the mare are given in Table 7-1. Mares with their first foal may not drip milk, or have noticeable udder development. Even in other mares there are times when all signs fail.

Most foals are born from 11 PM to 3 AM. When foaling is near, wrap the tail (or it may be easier to put it in a tube sock taped at the top around the tailhead) and wash down the perineal area. If the mare has had a Caslick's closure of the top of the vulva, snip it open with a pair of sharp, clean scissors.

THE ACT OF FOALING

Uterine contractions force the *placenta* through the cervix into the vagina; occasionally it may be seen protruding at the vulva. Following its rupture and the expulsion of several gallons of fluid, the mare may rest for 10 to 15 minutes before actual labor begins. The mare usually lies down during forceful contractions. However, about 10% of mares

---

*Words in italics are defined in the glossary.

TABLE 7–1
SIGNS OF APPROACHING PARTURITION IN THE MARE

Signs	Time Before Foaling
Distended udder	2–4 weeks
Shrinkage of buttocks near tailhead, and a drooping of the abdomen (more pronounced with age)	1–3 weeks
Filling out of the teats	4–6 days
End of nipples covered with a wax-like material	1–4 days
Loose vulva and milk drops	½–1 day
Restlessness, pacing, seeking isolation, breaking into sweat, frequent urination, lying down and getting up. It is important to let the mare do this, unless she is hurting herself, as it helps the foal get into position for birth.	2–3 hours
Ruptured membranes, and 2 to 5 gallons of fluid expelled	30–60 minutes
Visible labor begins	15–30 minutes
Expulsion of afterbirth	15–120 minutes after foaling

will foal standing. The *amnionic sac*, which is white and glistening, is the first thing presented. This is followed by the foal's feet inside the amnionic sac. They should be soles down, with one foot 2 to 6 inches in front of the other. If the distance is more than this, there is a chance that one elbow is locked on the brim of the mare's pelvis. The foal's nose should be near his *carpus* (Glossary Fig. 1); if it is not, it may indicate that the foal's head is turned back. If presentation is not normal, and just the feet are showing, get the mare up and walk her around. This may correct the foal's position. Once the feet are showing, there should be progress within 5 minutes. Normal uterine contractions and abdominal effort expel the fetus in approximately 15 minutes or less. If there is any variation from this, the owner should immediately call the veterinarian. During foaling, do not try to help the mare; stay away as much as possible, unless there is trouble.

As soon as the foal is born, make sure the *amnionic sac* is off its nose. This is one of the major reasons for supervising the birth. Wait 30 seconds, and if the foal doesn't breathe, rub it, hold its head down and shake it, and remove material from its mouth and nostrils. The foal will usually do this itself. If the foal is not breathing by 60 seconds following birth, inflate the lungs by blowing into one nostril while closing the opposite nostril and mouth. This initial inflation may be all that is needed to get respiration started. If it is not, establish a rhythm of about 25 times per minute, until the foal begins to breathe normally.

Failure to provide oxygen in the first two to three minutes of life will result in permanent brain damage or death. However, many problems are created or accentuated by well-meaning and concerned people who feel they must assist the mare in delivery, hasten the rupture of the umbilical cord, help the foal stand, and force the foal to nurse.

Try to leave the *umbilical* cord attached for several minutes, and let the mare lie still and rest. The foal may receive as much as 1500 ml of blood from the *placenta*. If the umbilical cord is broken too quickly, blood will squirt from the foal's umbilicus. To prevent this blood loss, the umbilical artery should be pinched closed with the fingers and held for several minutes. During this time, the artery constricts, and it can be released without any further loss of blood. If the umbilical cord hasn't broken by itself after 15 minutes, find the constriction in it, which is usually about 2 inches from the abdominal wall, and at this location twist and pull on it a little to break it. Then soak the stump for several minutes in a 7% tincture of iodine.

The *placenta*, or afterbirth, which normally weighs 10 to 12 pounds (4.5 to 5.5 kg), is usually expelled within 15 minutes to 1 hour following foaling. Occasional cramping pains, similar to mild *colic*, may occur during expulsion of the placenta and for a few hours following, as the mare's *uterus* continues to contract and involute. If the placenta is not passed by 4 hours following foaling, pull on it very gently. If it doesn't come out easily, veterinary medical treatment is recommended. Retained placenta in the mare may result in uterine infection (metritis), infertility, *founder*, or death.

CARE AFTER FOALING

The foal, if healthy, will stand within 15 minutes to three hours, and has a suckling reflex within 20 minutes. Don't try to get the foal up or help it walk. Just leave it alone and let it stumble around. Trying to help only exhausts the foal.

Prior to the foal's nursing, wash the mare's udder with a good disinfectant solution, rinse, and wipe dry. Normally, the foal will nurse within 1½ to 2 hours, although this may range from ½ to 6 hours. If the foal hasn't nursed by 6 hours, or is weak, milk some *colostrum* from the mare, or get some from a frozen source, and feed it to the foal from a nipple-bottle or stomach tube.

Colostrum is the first milk available to the foal after birth. It is secreted before, during, and shortly after foaling. It contains *antibodies* that are essential for the foal's resistance against *infectious* diseases. The quantity of antibodies present in the colostrum declines rapidly as milk is removed from the udder; the ability of the foal's intestines to absorb colostrum also declines rapidly during its first day of life. Therefore, as with all newborn animals, it is important for the foal to get adequate quantities of colostrum during the first day of life. If the mare has dripped milk for more than an hour or two prior to the foal's nursing, she may have lost most of her colostrum, and the foal may need to be fed mare's colostrum obtained from another source. Cow's colostrum is of little benefit in increasing the foal's resistance against infectious diseases, but is better than milk and should be used if mare's colostrum is not available.

If, for any reason, the foal doesn't receive normal quantities of its mother's *colostrum*, and it is 12 hours old or less, give the foal 1 pt (470 ml) of mare's colostrum by nipple-bottle or stomach tube every hour for three to four feedings. If the foal is over 18 hours old, it can no longer absorb the *antibodies* present in the colostrum. If it has not received adequate amounts of colostrum containing high antibody levels, as indicated by testing its blood and finding a *plasma* antibody or *immunoglobulin* level of less than 400 mg%, blood plasma from another horse should be given to the foal into the vein. A minimum dosage of 10 ml/lb body weight (22 ml/kg) should be given over a 1 to 2 hr period of time. At this dosage, most foals will need 1 L (1 qt) of plasma, which will increase their plasma antibody level to 30% of that of the donor.[59]

Plasma and *colostrum* from suitable donors, preferably from the same farm, can be frozen for future use. Every breeding farm or brood mare operation should maintain a bank of frozen colostrum. The colostrum can be obtained by milking 1 pt from each of several mares, shortly after foaling. Several quarts of colostrum in the freezer is good insurance for the time when a mare is not able to feed a newborn foal. Even when the mare is able to feed the foal, giving 1 pt of colostrum from the pooled source to the foal during the first several hours of life is an excellent procedure to follow. If kept frozen, colostrum may be stored for several years.[59] To prevent severe or even fatal *hemolytic* reactions, plasma or colostrum should be monitored for anti-*red blood cell isoantibodies*. If this cannot be done, pooled plasma from three to four male horses that have never been transfused, is an acceptable alternative.[59] One plasma transfusion is reported to be adequate.[59]

If the foal has not passed feces within 4 to 12 hours, an enema should be given. Since it is often not known whether feces have been passed, it is best to give an enema as soon as possible after birth, particularly to colts. Furthermore, an enema is often necessary for the elimination of the *meconium*, the sticky, tarry feces the foal accumulates before birth. An enema containing a surface-active agent, such as diocytl calcium sulfosuccinate (Surfak),* is recommended. Insert 4 oz, and hold the tail down for a few minutes to prevent prompt elimination. Don't use soapy water or *mineral* oil as an enema. If the enema is

---

*National Labs Corporation, Sommerville, NJ 08876.

not effective, have your veterinarian give mineral *oil* by stomach tube. If the mineral oil is not passed in the feces within 24 hours, call the veterinarian.

The foal is deficient in *vitamin A* when it is born. Unless the foal receives sufficient quantities of colostrum from a mare that has adequate vitamin A in her *ration* (e.g., is receiving good quality green hay or pasture), an injection of 300,000 IU of vitamin A should be given to the foal. (Both vitamins A and D are present in most injectable preparations.)

Immunization programs should begin at 6 to 12 weeks of age, and parasite control may be started as early as 6 weeks of age, depending on individual farm circumstances (see Vaccination and Worming, Chapter 10). Hoof trimming may be advisable in the first few months of life if normal hoof wear is not evident, or if corrective measures are needed. Whenever possible, put several foals together. This encourages playing and exercise, which is beneficial for normal growth and the development of a competitive spirit. Bone density is greater in animals that are highly active than in those that have only moderate or low activity.[99] Physical activity is required to develop and maintain a dense bone.[99] This speaks strongly against keeping growing horses of any age in stalls or small runs. Frequent handling, teaching to lead, and picking up all four feet, beginning from the first week of life is recommended.

HEMOLYTIC ICTERUS

Hemolytic icterus, or *neonatal isoerythrolysis,* has been estimated to affect 1 to 2% of all foals born. It is a disease similar to Rh blood type incompatibility in humans. It differs, however, in that horses have no Rh factor, but instead may be due to two or three other blood factors. It also differs from Rh factor disease in humans in that infants are affected prior to birth, whereas the foal is not affected until it nurses its mother's *colostrum.* The foal is not affected at birth because the mare's *antibodies* are not able to cross the *placenta* and reach the foal, whereas antibodies are able to cross the placenta in humans.

The disease is caused by the foal's absorption of *antibodies* produced by its dam and secreted into her *colostrum.* These antibodies destroy the foal's *red blood cells.* The destruction of red blood cells causes anemia and, if severe enough, *icterus* or *jaundice.* These antibodies do not harm the mare's red blood cells because she has a blood type compatible with them. However, if the foal inherits a blood type that is not compatible with these antibodies and absorbs these antibodies from the colostrum, the disease occurs.

Affected foals are normal at birth but within 1 to 3 days become progressively weak and lethargic. The heart rate and respiratory rate increase. When severely affected, foals may have blood in the urine, and *icterus.* Icterus may be detected earlier by taking a blood sample and observing the plasma after the red blood cells settle.

If the disease is detected before *icterus* is present and when symptoms and *anemia* are mild, the only treatment that may be necessary is to not allow the foal to nurse its dam for 36 hours and to keep the foal as quiet as possible.

Keep the foal in a stall and minimize all stress. However, if icterus is present and clinical signs and anemia are more severe, a blood transfusion is necessary. *Red blood cells* from the foal's mother should be given. Her red blood cells are not affected by the antibody causing the destruction of the foal's red blood cells and are therefore the best ones to give the foal. Several liters of the mare's blood should be taken. The red blood cells should be allowed to settle and as much of the *plasma* as possible siphoned off. The red blood cells should then be given intravenously to the foal.

A screening test to determine whether this condition will occur may be conducted by mixing 1 drop of blood from the foal's *umbilical stump* with 4 drops of 0.9% saline solution and 5 drops of the mare's *colostrum* on a clean glass slide. If true *agglutination* occurs within several minutes, the disease will occur if the foal ingests this colostrum. If the foal has nursed the mare before the agglutination test between the colostrum and the foal's blood is conducted, a false positive result may be obtained. In that case, an agglutination test can be run, using *serum* from the dam and *red blood cells* from the sire. Agglutination of the stallion's red blood cells in the presence of the dam's serum at dilutions of one part red blood cells, or greater, to two parts serum indicates *hemolytic icterus* may occur in the foal if it receives its dam's colostrum. This test may also be run before foaling, if possible imcompatibility is suspected.

It is frequently difficult to interpret an *agglutination* test (either between the *colostrum* and foal's blood or between the mare's *serum* and *red blood cells* from the sire). In addition, the disease, although most commonly caused by anti-red blood cell antibodies that cause red blood cell agglutination, may be caused by *antibodies* that result in *red blood cell* lysis instead, and therefore will not be detected using the agglutination tests described. A much surer means of preventing the disease is to have a *serum* sample, taken from the mare three to four weeks before foaling is due, checked for anti-red blood cell antibodies. Her serum should be checked for both agglutinins and lysins. Because of the low incidence of the disease, this is not practical for most mares. However, it may be for the mare that has previously had an affected foal, since some or all of her subsequent foals may be affected. Even for this mare, it may be more practical not to check her serum or to conduct either agglutination test, but instead, not to allow the foal to nurse the mare for the first 36 hours of life.

If *agglutination* occurs when blood from the foal and its dam's *colostrum* are mixed together, there is agglutination of the sire's blood when mixed with the dam's *serum*, agglutinins or lysins are present in the mare's serum, or the foal is from a mare that has had previous foals that were affected, the foal should be muzzled to prevent it from nursing its mother for the first 36 hours of life. One pint (470 ml) of colostrum from another mare should be given every 1 to 2 hours for three to four feedings, then a milk replacer should be fed from a nipple bottle, until the foal is 36 hours old (see Chapter 8). After this age, it may be allowed to nurse the mare and be raised in the normal manner. Ideally, the colostrum given should be checked with blood from the foal to ensure that agglutination does not occur. If it does, a different colostrum should be used.

Although it is frequently recommended that the dam be milked until the foal is allowed to nurse, to flush *colostrum* from her udder, this is not necessary other than to relieve pressure on the udder. After 36 hours of age, the foal is no longer able to absorb colostral *antibodies*; therefore, even if anti-*red blood cell* antibodies are ingested after this age, they are no longer harmful.

Chapter 8
Feeding the Young and Growing Horse

In feeding the growing horse from nursing to maturity, make sure that *free-choice** *trace-mineralized* salt and all of the good quality water and *forage* the horse can consume are available. During growth, good quality alfalfa, or other *legumes*, are preferred over grass forage. Compared with equal quality forages, legumes contain about two to three times more *protein*, three to six times more calcium, *beta-carotene* (the *vitamin A* precursor), and vitamin D, an equal amount of phosphorus, and several times more *lysine* than grass hays (Table 1–4 and Appendix Table 1). As shown in Table 8–1, all of these *nutrients* are needed in increased quantities for growth. However, if a good quality legume forage is not available, a grass *roughage* may be fed, and additional amounts of protein and calcium *supplements* added to the *concentrate* mix.

None of the feeds contain adequate phosphorus, and grass *roughages* and *cereal grains* do not contain adequate *protein* or calcium to meet requirements for growth (Table 8–1). Therefore, a *concentrate* mix must be fed that contains sufficient quantities of protein, calcium, and phosphorus *supplements* to make up the difference between that needed and that present in the roughage. Concentrate mixes of this type are given in Appendix Table 6, or can be formulated as described in Chapter 3. High *lysine*-containing protein supplements, such as soybean meal or animal-source protein supplements (Table 1–4), should always be used for the growing horse because of the high level of lysine necessary for growth (see Chapter 1).

Separate feeding programs are necessary to meet the nutritional requirements of nursing foals, weanlings, and yearlings until they reach 90% of their mature weight (Appendix Table 7).

---

*Words in italics are defined in the glossary.

TABLE 8–1
NUTRIENTS NEEDED BY THE GROWING HORSE AS COMPARED TO THOSE IN FEEDS[50]

Required or present in the feed*	Protein	Calcium	Phosphorus	Ca : P ratio
Requirements:	——% in total air dry ration——			
Creep feed	16	0.80	0.55	1:1 to 3:1
Weanling	14.5	0.65	0.45	1:1 to 3:1
Yearling	12	0.50	0.35	1:1 to 3:1
Composition of:*	——% in air dry feed——			
Alfalfa (or other legumes)	13–19	0.80–2.00	0.10–0.30	3:1 to 10:1
Grass	5–13	0.30–0.50	0.10–0.30	0.8:1 to 1.4:1
Cereal grains	9–12	0.02–0.10	0.25–0.35	0.05:1 to 0.3:1

*For more exact values, see Appendix Table 1 for the specific type of grain or roughage being fed, and for the most accurate values, have the feed analyzed.

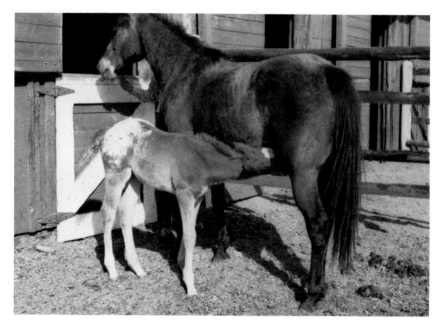

Fig. 8–1. The foal's nutritional requirements are completely met by the mare's milk during the first two months of life, provided that the mare produces a normal amount of milk.

THE NURSING FOAL

The mare's milk (Fig. 8–1) plus the pasture, hay, and *concentrate* mix being fed the mare, which most foals will start nibbling by a couple of months of age (Fig. 8–2), is sufficient to meet the foal's nutritional requirements during the first three months of life. An exception to this is when the mare is not producing a normal amount of milk. This is indicated by a thin foal.

Occasionally a mare will produce adequate quantities of milk, but with improper amounts of either calcium or phosphorus, or less commonly, other *nutrients*. An imbalance of calcium or phosphorus in the milk is unrelated to the mare's diet, and cannot be changed by altering the *ration;* it is inherent in that mare. Dietary manipulation and feed intake affect the quantity of milk produced, but have a

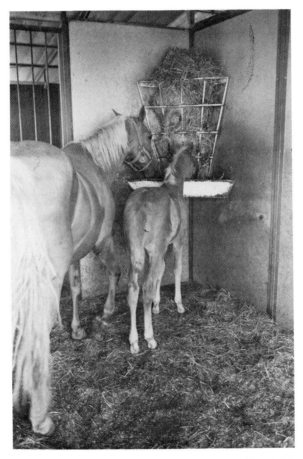

Fig. 8– 2. Most foals at an early age will begin eating the grain and hay fed the mare, and do not need to be fed a creep ration until they are 3 months old.

minimal effect on the nutrient composition of milk.[72] The *mineral* imbalance in the milk may result in *epiphysitis*, causing enlarged joints and crooked legs, or in a decreased growth rate. If a normal, nursing foal, less than 3 months of age, develops epiphysitis, an imbalance in the mare's milk or an excessive amount of milk production should be suspected. An imbalance in the mare's milk can be confirmed by having the milk analyzed. It should contain the nutrient contents given in Table 8–2. If it does, then the mare may be producing more milk than the foal can tolerate, resulting in either a loose, pasty stool or epiphysitis or both. To correct this, reduce the mare's feed in order to decrease her milk production.

If the quantity of *nutrients* in the mare's milk is inadequate, particularly in calcium or phosphorus, the foal should be raised as an orphan as described later in this chapter. After five weeks of age, the foal need not be treated as an orphan, and should be fed 1 to 1.5 lbs/100 lbs body wt/day/ (1 to 1.5 kg/100 kg/day) of a *creep feed* and good quality, *free-choice roughage*. A creep feed should contain the nutrient levels given in Table 8–1. Creep feed may be mixed as given in Appendix Table 6 (mix No. 2), formulated as described in Chapter 3, or bought already mixed. At three months of age, non-nursing foals may be fed as weanlings.

If the foal and mare are doing well, it is best not to feed the foal *concentrates* separately from the mare until it is at least two months, and preferably three months, old. After the third month of lactation, the mare's milk production declines while the foal's nutritional needs continue to increase. The difference between the *nutrients* received from the milk and *forage*, and those needed, should be provided by feeding a *creep feed*. Feed 0.5 to 0.75 lbs/100 lbs body wt/day of a creep feed consisting entirely of concentrates (0.5 to 0.75 kg/100 kg/day).

If the foal is not fed individually, have the *concentrates* used in the *creep feed* mixed with 50% chopped *roughage* to limit concentrate intake. This mixture may then be offered *free-choice* (Fig. 8–3). Free-choice feeding of creep feeds consisting entirely of concentrates is not recommended, since some foals may consume excessive amounts of concentrate, which can result in *epiphysitis* (see Chapter 9) or enterotoxemia. Enterotoxemia usually results in death within a few hours, preceded by colic and labored breathing. It is caused by the proliferation of the bacteria Clostridium perfringens Type D in the small intestine. Although the free-choice feeding of creep feeds consisting entirely of concentrates does not in many instances result in epiphysitis or enterotoxemia, it is safer to limit the intake of creep feed consisting entirely of concentrates to the amounts described in the preceding paragraph. This does not, however, guarantee against excessive concentrate consumption when feeding several foals in a

TABLE 8–2
COMPOSITION OF MARE'S AND COW'S MILK

	Total Solids (%)	Protein (%)	Fat (%)	Lactose (%)	Ca (mg%)	P (mg%)	Mg (mg%)	Se (ppm)	Cu (ppm)	Zn (ppm)	Fe (ppm)	Mn (ppm)
Mare's milk at foaling	25	19	0.7	5	80–120	45–90	6–12	0.01–0.03	0.8–1.2	5–8	1.0–1.6	0.5
Mare's milk 2 mo after foaling	10*	2*	1.5	6					0.15–0.4	2–4	0.5–0.9	0.25
Cow's milk	13	3.5	3.7	5	95–170	70–160	8–27					0.5

*The decrease in the content of protein and total solids occurs during the first 12 hours of lactation. Most, although not all, the decline in mineral concentration also occurs during the first 12 hours of lactation.

Fig. 8–3(A,B,C). Creep feeders. C shows the inside of the creep feeder in B. Note that the openings into the feeders are narrow enough so the mare can't get in, but the foal can. A feed box containing bars across the top placed close enough together to prevent the mare from inserting her nose into the box but far enough apart to allow the foal to eat from the box may also be used as a creep feeder.

138

group, since dominant foals may receive excessive amounts, and more timid foals receive less. Watch the foals; if this is occurring, remove either the more dominant or the more timid foals from the others, and feed them separately.

Weaning at about four months of age is preferred. This prevents excessive loss of condition by the mare. Both the amount of milk produced by the mare and its *nutrient* content are decreased by five months of lactation. By this age, the foal receives little of its nutritional needs from the mare's milk. In addition, it is more difficult to properly feed the older foal while it is still with the mare. However, weaning prior to three months of age is not routinely recommended. Although it has been shown that some foals can be weaned at two months of age, and do as well as those weaned at four months,[17] early weaning necessitates feeding an increased amount of *concentrate* at a younger age. This may result in the overconsumption of concentrate, which in some foals can cause, or predispose them to, *epiphysitis.*

The weaned foal may be turned out on pasture with other weanlings two days after weaning. If there are numerous weanlings, separate them by sex and temperament. Weanlings should not be run with older horses.

THE WEANLING

The weaned foal should be fed 1 to 1.5 lbs/100 lbs of body wt daily (1 to 1.5 kg/100 kg/day) of a *concentrate* mix similar to those given in Appendix Table 6, or a mix that can be formulated as described in Chapter 3 to meet the weanling's requirements (Table 8–1). The weanling should consume at least 1 lb/100 lbs body wt daily (1 kg/100 kg/day) of good quality *roughage.* If this amount is not being consumed, the quantity of concentrates available should be reduced until this amount of roughage is consumed. Feeding high levels of concentrates will cause the weanling to grow more rapidly. However, excessive growth rates may predispose the animal to bone and tendon problems. Therefore, overfeeding concentrates should be avoided. A horse growing at a normal rate will be a better performer, and a more durable athlete than a horse fed to accelerate the rate of growth. In addition, there will be no difference in size at maturity.

THE YEARLING

By 12 months of age, the horse's growth rate has slowed considerably (Appendix Table 7), and there is a corresponding decrease in its nutritional requirements per pound of body weight. Therefore, the amount of *grain* fed to the yearling should be decreased to 0.5 to 1

TABLE 8–3
AMOUNT OF CONCENTRATE TO FEED FOR OPTIMUM GROWTH RATE*

	Amount of Concentrate (lbs/100 lbs body wt/day or kg/100 kgs body wt/day)
Nursing foal	0.5–0.75
Weanling	1.0–1.5
Yearling to 90% of mature wt	0.5–1.0

*To be fed along with all the roughage the horse will eat. Feeding more concentrate than this may cause epiphysitis and contracted flexor tendons in some horses (see Chapter 9). Feeding less concentrate will result in a slower growth rate. A good thumb rule to follow is to feed each foal 1 lb of concentrate per day for each month of age up to a maximum of 8 to 9 lbs per day. Continue feeding this amount until 90% of anticipated mature weight is reached.

lb/100 lbs of body wt daily (0.5 to 1 kg/100 kg/day) until 90% of mature weight is attained (Appendix Table 7), at which time, it should be fed the same as the mature horse for maintenance or work. The amount of *concentrate* needed to obtain maximum growth rate, with minimum chance of bone and tendon problems, is given in Table 8–3.

RAISING ORPHAN AND EARLY-WEANED FOALS

If the foal is orphaned at or near birth, the same procedures described in Chapter 7, under Care After Foaling, are necessary. If a foal's mother is unable to raise it, the foal may be "grafted" onto another lactating mare, if one is available. It may take up to 10 days before a foster mare will show maternal behavior toward the foal, but if the mare can be physically restrained or chemically tranquilized so that she will not injure the foal, she should eventually accept it.[35] Mares use all three senses—smell, sight, and sound—to identify their foals.[35] Smell is used for close range identification. Therefore, when trying to get a lactating mare to adopt a foal, it is helpful to make the foal smell like the mare by coating it with her sweat, milk, or feces, or by temporarily masking the mare's sense of smell by smearing something like Vicks in her nostrils.

If the foal is to be hand-raised, bucket- rather than nipple-feeding is preferred, because feeding equipment is easier to keep clean, and feeding is much easier and faster. Plastic buckets with relatively wide openings, rather than tall, narrow, or metal buckets are best. To start the foal drinking, let it suck on a finger. If the foal does not nurse the finger, move the finger against the palate and the tongue to trigger the foal's nursing response. As the foal sucks on the finger, slowly and gently move it, and the foal's mouth, down into the milk until it begins drawing milk from the bucket. The finger is then removed. This is the point at which patience is very important. The foal may fight at first because it does not understand what is being done. After several attempts of this sort, however, the foal will learn to drink from the bucket. The bucket containing milk can be hung in a convenient location, and left for the foal. It is best to use a light-colored bucket so the foal can see it easily. There is no benefit in warming the milk before feeding. Cold milk is preferred because its temperature remains more constant.

A good quality mare's milk-replacer should be fed. However, a calf milk-replacer is satisfactory, providing it contains iron, and is good quality (i.e., contains at least 20% crude *protein*, and 15% *fat*, and no more than 0.5% crude *fiber*). These may be fed in the amounts and at the frequencies given in Table 8–4. However, milk may be left with the foal at all times, without fear of its overeating. When this regimen is followed, foals will drink small amounts of milk periodically during the day and night. The milk should be changed and replenished twice a day. Each time the milk is changed, the bucket should be thoroughly cleaned.

When the foal is several days old, milk-replacer pellets (e.g., Foal-Lac or Mare-Lac pellets, see Table 8–4) should be placed in its mouth several times a day, and as much as it will eat placed in a bucket in the stall. Any pellets remaining should be discarded twice a day, and new pellets put in the bucket. When the foal is eating 2 to 3 lbs (0.9 to 1.4 kg) of milk-replacer pellets daily, the pellets can be gradually replaced with a good quality *creep feed* containing 16 to

TABLE 8–4
FEEDING THE ORPHAN FOAL

Age	8 oz Cups of Foal-Lac or Mare-Lac*/Feeding	Pints of Water/Feeding	Times/day to Feed
0 to 1 day (feed 1 pt, or 470 mls, of colostrum)	—	—	4
1 to 5 days	1 (¼ kg)	1½ (750 ml)	4
6 days to 4th or 5th week	2 (½ kg)	3 (1500 ml)	3

Second week—feed Foal-Lac or Mare-Lac pellets*, or a creep ration.

Third week—the foal should be eating ¼ lb (0.13 kg) of the pellets or creep ration daily; increase this to 2 lbs (1 kg) daily by the fifth week; then wean.

Fifth to twelfth week—feed ¾ lb of creep feed/100 lbs body weight (¾ kg/100 kg/day) and alfalfa.

3 to 12 months—provide a weanling ration (see earlier in this chapter).

*Foal-Lac (Borden Chemical, Box 419, Norfolk, VA 23501) or Mare-Lac (Diamond Labs, P.O. Box 863, Des Moines, IA 50304). If these are not immediately available, use a good quality calf-milk replacer (one that contains 0.5% crude fiber or less, 15% or more fat, and 20% or more protein), until a mare's milk replacer can be obtained.

20% protein. The feed bucket should be cleaned and filled with fresh feed daily. By four to five weeks of age, the foal should be eating at least 2 lbs (1 kg) of *creep ration* daily, at which time milk feeding can be stopped. Excellent quality hay or pasture should be made available to the foal at this time.

When good quality alfalfa hay or pasture is available to an orphan foal of less than one or two months of age, it may overeat and develop diarrhea. At such an early age, the foal is unable to properly utilize large amounts of *forage* or *fiber.* If this occurs, the amount of forage available must re restricted.

Chapter **9**

Epiphysitis and Contracted Flexor Tendons in the Growing Horse

CLINICAL SIGNS

The syndrome called *epiphysitis** occurs most commonly in the rapidly growing horse. It is characterized by flaring and enlargement of the *metaphysis* of the long bones. The name epiphysitis is a misnomer, since the inflammation involves primarily the metaphysis, and not the epiphysis, and therefore metaphysitis would be more correct. However, epiphysitis is the name commonly used to describe this condition, and therefore will be used in this text. The condition occurs most commonly at the distal *radius*, but may be noticeable at the distal *tibial*, and distal *metacarpal* and *metatarsal* bones (see Glossary, Fig. 1). Swelling occurs just above the *carpal joints*, and gives them a dished-in appearance in front, which is sometimes referred to as *"open knees"* (Figs. 9–1 and 9–2). Epiphysitis occurring at the metacarpal and metatarsal metaphysis causes enlarged *fetlocks* (bumps) (Fig. 9–2). Some horses may have angular deformities of either, or both, carpi or fetlocks (Fig. 9–1).

Contracted *flexor tendons* may be present at birth, or they may be acquired.[46,57] They can be accompanied by angular limb deformities, with signs of *epiphysitis*. Contracture present at birth as a result of malformation of the joints, bones, or both is generally untreatable. If it is due to *intrauterine* malpositioning, it can be treated with varying degrees of success, depending on the type and severity of the condition. Treatment may require using splints or casts, although these aren't always necessary.[57] Acquired tendon contracture occurs during growth, and is first characterized by the *pasterns* being more

---

*Words in italics are defined in the glossary.

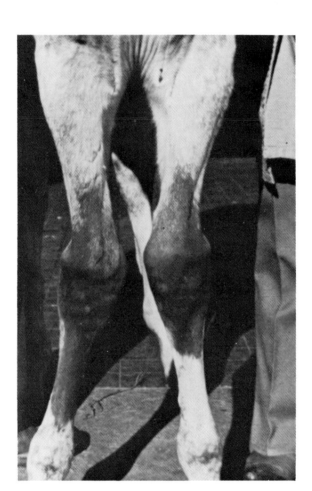

Fig. 9–1. Medial deviation of the foal's right carpus, and "open knees" in both legs as a result of epiphysitis.

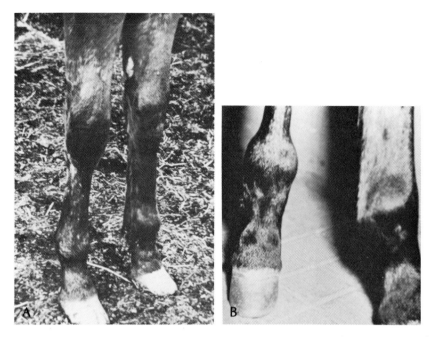

Fig. 9–2(A,B).   Acquired contraction of the flexor tendons causing straight pasterns (A), and epiphysitis causing "open knees" and enlarged fetlocks (B).

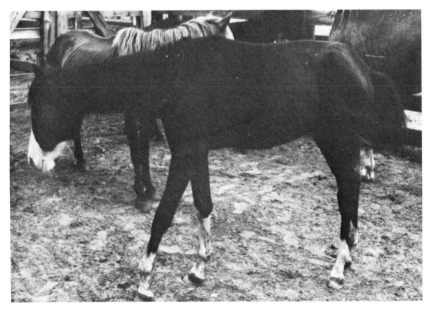

Fig. 9–3.   Acquired contracted flexor tendons in a horse receiving inadequate quantities of feed. Note how straight up and down the fetlocks and pasterns are, and note the knuckling forward of the left front fetlock joint. Although contracted flexor tendons occur more commonly in the growing horse receiving excess energy, it may also occur as a result of inadequate feed intake. This may be due primarily to a protein deficiency.

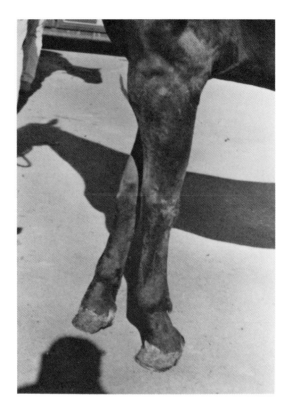

Fig. 9-4. Acquired contracted flexor tendons in a yearling being fed excessive amounts of grain.

nearly vertical than normal (Fig. 9-2). More severe cases may knuckle forward at the *fetlock* joint, causing the horse to walk on the toe (Figs. 9-3 and 9-4). Epiphysitis is thought to be responsible for causing acquired contracted flexor tendons by creating a neurogenic stimulus for contracture as a result of the pain associated with metaphyseal inflammation. This is the basis for the use of *analgesics* and anti-inflammatory drugs in the treatment of contracted flexor tendons.

CAUSES

The major factors that cause or predispose the animal to *epiphysitis*, acquired contracted *flexor tendons*, and angular limb deformities are (1) rapid growth, (2) trauma to the *epiphysis*, (3) nutritional imbalances, and (4) genetic predisposition. These four factors are interrelated, and there are a number of additional factors that affect each.

Rapid growth rate is promoted by (1) the animal's genetic capacity, (2) high *energy* intake, and (3) stunting early in life, followed by increased feeding for maximum growth.[37,74] The greater the animal's genetic capacity for growth and the larger the body size, the greater the susceptibility to these conditions.[25]

Trauma to the *epiphysis* causes pain due to epiphyseal inflammation *(epiphysitis)*.[57] The opposite leg may subsequently become affected, due to its increased bearing of weight. Epiphyseal trauma increases as the weight per unit of cross-sectional area at the epiphysis increases. The greater the body weight, as a result of either genetics or feeding, and the smaller the bone, the more susceptible the animal is to this condition.[57] Therefore, epiphyseal trauma and growth rate are related as causative or predisposing factors in epiphysitis. This condition occurs more commonly in the faster growing, larger, finer-boned breeds of horses, such as Thoroughbreds and modern-day Quarter Horses. The condition is less common in draft horses because they have larger bones and a slower rate of growth (Appendix Table 7), and in ponies because they have a smaller body size. Although the condition may occur in either the front or back legs, it is much more common in the front because 60 to 65% of the horse's weight is supported by the front legs. Thus, trauma to the epiphysis, or weight per unit of cross-sectional area at the epiphysis, is nearly twice as great in the front legs as it is in the back legs.

Lack of exercise has been implicated as a cause or predisposing factor of *epiphysitis*.[52] Exercise increases the breaking strength of the bone[65] and has a *hypertrophic* effect[46] on long bones in growing horses. Thus, the growing horse should be allowed unlimited exercise; it should be kept from shortly after birth in an area large enough to permit unlimited running, bucking, and playing. Having two or more foals in the same pasture is also beneficial in encouraging playing, and in the development of a competitive spirit.

Several nutritional imbalances may predispose the animal to, or cause, *epiphysitis*. These include calcium and phosphorus imbalances, inadequate *protein*, excessive *energy*, or any combination of these factors.[6] Adequate quantities of calcium and phosphorus must be present in the *ration* to meet the requirements of the growing horse (Appendix Table 2). The amount of calcium with respect to phosphorus is called the *Ca:P ratio*. It may vary from 1 to 3 parts calcium, to 1 part phosphorus. Epiphysitis may occur when the Ca:P ratio is out of this range, or the amounts of either are inadequate. Excess calcium and phosphorus intake may also be detrimental for optimal development and maintenance of bone. In mature mice a dietary calcium and phosphorus concentration four-fold over that required decreased bone thickness and mineral content even though the diet contained a Ca:P ratio of 2.[100]

Nutrients other than calcium and phosphorus are necessary for proper bone growth.[34] The nutrient most commonly deficient is *protein*, and the ones most commonly in excess are those that provide *energy*. Since protein constitutes 20% of the bone matrix,[34] inadequate protein intake is detrimental to proper bone growth and development (Fig. 9–3). Excessive protein has also been implicated as a causative factor in *epiphysitis* and contracted *flexor tendons*.[1] However, this has not been confirmed. One explanation of this is that protein is also a source of energy, and therefore, an excessive intake may result in an intake of excessive energy, which has been shown to be an important predisposing factor to epiphysitis. In addition, urinary calcium excretion may increase with increasing levels of dietary protein. In one study in humans, an 800% increase in urinary calcium excretion occurred when protein was increased from 6 to 600 g per day, and the relationship was linear over this range.[44] If this occurs in the horse, excessive protein intake may cause a calcium deficiency and result in epiphysitis, or other skeletal problems. Increasing dietary calcium will reduce, but not overcome, the calcium imbalance induced by the consumption of high-protein diets.[43] Thus, protein intake should be monitored, so that it does not greatly exceed the animal's requirements (Appendix Table 2).

The most probable contributory effect of *protein* intake on *epiphysitis* and contracted *flexor tendons* is its effect on the growth rate of the horse. Feeding protein above the animal's requirement does **not** increase growth rate above that achieved when the *ration* just meets the animal's protein requirements as given in Appendix Table 2. However, inadequate protein intake slows growth rate, regardless of whether there is adequate *energy* available, so that increasing the protein content of a previously protein-deficient, but energy sufficient, ration to that necessary to meet the animal's requirement, permits a more rapid growth rate. If the ration does not contain adequate *minerals*, such as calcium or phosphorus, to support a faster rate of growth, epiphysitis, and contracted flexor tendons may occur. It appears that increasing the protein content of the ration causes these conditions, and that decreasing the protein content of the ration prevents them. In reality, the level of protein in the diet simply controls growth rate. The cause of the conditions is inadequate calcium, phosphorus, or other minerals in the ration. The amount of protein in the ration simply masks the deficiency by slowing growth rate, or unmasks it by permitting a faster growth rate to occur. The same may be true for the role of energy intake in causing epiphysitis or contracted flexor tendons.

In several species, including the horse, excessive *energy* intake during growth has been shown to increase abnormal bone conditions, such as lower specific gravities and cortical thickness, decreased ash content, *osteochondritis*, and *epiphysitis*.[24,31,38,45,60] Maximum energy intake and growth rate appear to be incompatible with optimal skeletal development. Although diets may be adequate for normal rates of growth, they may not be adequate in some unknown *nutrient(s)* for maximal growth rates. Therefore, to prevent abnormal bone conditions, energy intake and growth rate must be controlled.

There may also be a genetic predisposition to *epiphysitis* and contracted *flexor tendons* that is unrelated to growth rate or bone size, as has been suggested in dogs[51] and pigs.[23]

The major factors responsible for *epiphysitis* and acquired contracted *flexor tendons* and angular limb deformities appear to be (1) rapid growth, (2) trauma to the *epiphysis*, (3) nutritional imbalances, and (4) genetic predisposition. It has been suggested that these factors are also contributing causes of *"wobbler's syndrome"* and *osteochondritis dissecans.*[45] A combination of two or more of these factors would increase the incidence and severity of these conditions, although any one of the factors may be responsible. For example, if a growing animal is forced to bear an excessive amount of weight on a limb, due to a fracture or painful soft tissue injury, epiphysitis, deviation and or contracture in the opposite limb may occur due to trauma to the *epiphysis.*[57] Periods of less than one week of excessive weight bearing on a limb in the young horse may lead to an angular deformity in that limb. The same thing may be produced by tying up one of the forelimbs of a growing animal. When this is done, the weight per unit of cross-sectional area, or trauma, to the other leg is increased and epiphysitis will occur.[24] On the other hand, calcium and phosphorus imbalances (deficiencies or excesses), inadequate *protein*, or excessive *energy* in the *diet* may result in epiphysitis, regardless of epiphyseal trauma.

DIAGNOSIS OF NUTRITIONAL CAUSES

The four dietary calcium and phosphorus imbalances that occur are (1) calcium excess, (2) calcium deficiency, (3) phosphorus excess, and (4) phosphorus deficiency. All result in similar signs of *epiphysitis*, which cannot be differentiated clinically or radiographically. *Radiographs* provide evidence of bony changes, but are of no value in diagnosing the cause of the condition. In addition, radiographs do not show bony changes until they are extensive and generally apparent clinically.[34] Radiographic signs of epiphysitis may include alterations in the axis of the limb, lipping of the *metaphysis*, and metaphyseal *sclerosis.*

Concentrations of *nutrients* in the hair, blood *plasma*, and urine, as will be described, are used to determine whether a nutritional imbalance is present. However, **the only accurate and reliable means of diagnosing dietary calcium excess, phosphorus deficiency, protein** excess or deficiency or energy excess, is to evaluate the ration. Excessive *energy* intake is present if more *grain* mix is fed than the amounts given in Table 8–3. A deficiency is present if the total *ration (concentrates* plus *roughage)* does not meet the calcium, phosphorus, and *protein* requirements (Appendix Table 2). For the growing horse, excessive dietary calcium or phosphorus is present when the *Ca:P ratio* is outside the range of 1:1 to 3:1. The nutrient content of the total ration for any horse may be determined and balanced as described in Chapter 3.

Concentrations of Calcium and Phosphorus in Hair and Blood Plasma

Analysis of hair for calcium and phosphorus is of no value in determining dietary calcium or phosphorus imbalances. There is no relationship between the amount of calcium and phosphorus in the hair and the amounts present in the *diet*. Further, there is a seasonal variation in the mineral content of the hair. It is lowest in the winter.[76]

Determining the concentrations of calcium and phosphorus in the *plasma* is of little benefit, and may be misleading in diagnosing dietary imbalances of these *minerals*. These values are useful only in diagnosing dysfunctions in the body's *homeostatic* mechanisms for calcium and phosphorus, such as *parathyroid hormone, calcitonin* secretion, or *renal* dysfunction. Nutritional imbalances result in no change or in only small cyclic changes in the concentrations of calcium and phosphorus in the plasma. For example, when phosphorus absorption is excessive and calcium absorption is inadequate, initially the concentration of calcium in the plasma ($[Ca]_p$) decreases and the concentration of phosphorus in the plasma ($[P]_p$) increases. At this time, the concentrations in the plasma are indicative of the nutritional imbalance. The decreased $[CA]_p$ stimulates increased parathyroid hormone (PTH) secretion. This hormone causes an increase in intestinal absorption and renal tubular reabsorption of calcium, calcium and phosphorus mobilization from the bone, and increased urinary phosphate excretion.[10] These effects return both $[Ca]_p$ and $[P]_p$ to normal, where they may remain. Often, there is overcompensation which increases the $[Ca]_p$ above normal, at which time the concentration in the plasma is exactly the opposite of the nutritional imbalance. The increased $[Ca]_p$ causes a decrease in PTH secretion. This returns $[Ca]_p$ to normal where the cycle may begin again. The amount of change generally observed is 1 to 2 mg% above or below a normal of 11 to 12 mg% of calcium, and during peak growth of the animal, 5 to 7 mg% of phosphorus. The $[P]_p$ is normally about 2 mg% higher during peak growth than in the mature animal.[36]

Renal Clearance Ratios

Renal phosphorus excretion is the most important mechanism for maintaining phosphorus *homeostasis*, and is directly related to phosphorus intake.[61] Therefore, the *renal clearance ratio* of phosphorus is helpful in diagnosing a relative, or absolute, phosphorus excess. The renal clearance ratio of phosphorus may also be helpful in diagnosing a deficiency of calcium, because if there is inadequate calcium for phosphorus utilization, there is an increase in phosphorus excretion in the urine. The percent renal clearance ratio of a substance (% C of X) is determined by measuring the *plasma* and the urine concentration of that substance ([X]) and of *creatinine* ([Cr]), and calculating its percent clearance ratio:[70,72] % C of X = $([X]_u/[X]_p) \times ([Cr]_p/[Cr]_u) \times (100)$.

The urine sample must be collected without the use of *diuretics* and within a few hours of when the blood sample was taken. Mucus excretion in the urine has no effect on phosphate clearance.[7] The normal *renal clearance ratio* of phosphorus is 0 to 0.5%, but may go as high as 1.0%.[7] Levels greater than this may be indicative of *hyperparathyroidism, renal* tubular insufficiency, or relative or absolute excess phosphorus intake. The phosphorus intake may or may not exceed the animal's requirement; it may simply be in excess of the amount of calcium available for phosphorus utilization. This results in increased excretion of phosphorus. Thus, an increased renal clearance ratio of phosphorus may also indicate a dietary calcium deficiency or a *Ca:P ratio* of less than 0.8:1. Since the renal clearance ratio of phosphorus may normally be zero, it is of no benefit in diagnosing a dietary phosphorus deficiency. Although renal excretion of calcium is important in maintaining calcium homeostasis,[63] renal clearance of calcium is of no benefit in diagnosing dietary calcium excess or deficiency. Urinary calcium excretion by the horse is variable, and appears to be unrelated to the *diet*.[7]

EFFECT AND TREATMENT OF NUTRITIONAL CAUSES

The major feeding practices responsible for nutritional imbalances in the growing horse are (1) feeding too much *grain*, (2) feeding a grain mix that is inadequate in calcium, phosphorus, and *protein* when a grass *roughage* is fed, and inadequate in phosphorus when alfalfa or other *legumes* are fed, and (3) feeding inadequate quantities of grain, which generally occurs as a result of a strictly roughage *ration*.

Excessive *concentrates* are most commonly fed in two situations: (1) allowing the foal continual access to a *creep ration* that does not contain *roughage*, and (2) feeding concentrates in amounts greater than those shown in Table 8–3. Although *cereal grains* contain inadequate amounts of phosphorus for the growing horse, they contain 3 to 20 times more phosphorus than calcium, i.e., cereal grains have a $Ca:P$ *ratio* 0.05 : 1 to 0.3 : 1 (Table 8–1). In addition to a low calcium content, cereal grains contain about 1% *phytate*.[28] The high phytate and low Ca : P ratio decrease calcium absorption.[28,62,63] Thus, the major effect of excessive *grain* intake is a calcium deficiency and a relative, although not absolute, excess of phosphorus. Cereal grains are also high in *energy*. They contain nearly twice the available energy as roughages. Thus, a *ration* high in cereal grains is high in energy, which promotes a rapid growth rate, but is inadequate in calcium and phosphorus, which are necessary for proper bone growth. The result may be *epiphysitis,* contracted *flexor tendons,* or angular limb deformities.

To prevent *epiphysitis,* feed the growing horse a properly formulated *concentrate* mix (Appendix Table 6), but not in excess of the amounts given in Table 8–3. If the condition occurs (and it may in some horses even with proper feeding), immediately stop feeding **all** *concentrates.* This decreases *energy* intake, slows growth rate, and allows increased amounts of calcium and an appropriate amount of phosphorus to be absorbed. Feed as much good quality grass hay as the horse will eat, feed no *grain*, and allow free access to a commercial salt-mineral mix that contains roughly equal amounts of calcium and phosphorus (Appendix Table 5), or to a mix made with equal parts of calcium phosphate and *trace-mineralized* salt. No other salt should be available. This will force the horse to eat what is provided. The loose, rather than the block, form of the salt-mineral mix is preferred because of higher consumption of the loose form. If grass hay is not available, alfalfa may be fed. In this case, a salt-mineral mix higher in phosphorus than in calcium is preferred (see Chapter 5 and Appendix Table 5). The horse should be eating 3 to 4 oz (85 to 115 g), or more, daily of the salt-mineral mix. If this amount is not being consumed, feed 1 lb (0.45 kg) of *sweet feed* or grain moistened with water, with 2 to 3 oz (55 to 85 g) of the salt-mineral mix added to it, twice a day.

Exercise appears to be important for normal bone growth.[46,52,65] However, once *epiphysitis* has occurred, exercise is detrimental, especially if a concurrent angular limb deformity is present. Exercise will cause greater trauma to the *epiphysis,* and increase the severity of the condition. Many mild angular limb deformities will return to normal with stall rest. Complete (24-hour-a-day) stall rest is recommended. The problem arises when contracted *flexor tendons* also occur with epiphysitis and angular limb deformity. Exercise may be beneficial for the contracted tendons, but increases the trauma to the epiphysis, increasing the severity of the epiphysitis and deformity.

There is controversy concerning the use of *analgesics* and anti-inflammatory drugs in the treatment of these conditions. They should not be used in the horse in which *epiphysitis* and angular limb deformity are the major problems. Their use decreases pain and inflammation, resulting in increased physical activity, which in turn increases the trauma to the *epiphysis*, increasing the severity of the condition, or at least slowing recovery. However, for contracted *flexor tendons* relief of pain is indicated, so analgesic drugs such as Butazolidin should be used.[57] Corticosteroids should not be given. Corticosteroids given daily cause extensive skeletal damage in the growing horse.[101]

Epiphysitis and angular limb deformities generally show improvement after 4 to 6 weeks of complete stall rest while feeding only *roughage* and the appropriate salt-*mineral* mix. Even if recovery is not complete after this length of time, begin feeding 0.5 lbs/100 lbs body wt/day (0.5 kg/100 kg/day) of a properly formulated commercial *concentrate* mix (Appendix Table 6), or one formulated as described in Chapter 3.

This management scheme is usually successful, but some horses may not respond to it. It is usually desirable to return the affected limb to normal alignment as soon as possible, to prevent secondary degenerative changes that may develop.[46] In cases where conservative treatment has failed, or in severe cases (greater than 15 degrees angulation), surgery may be necessary. For acquired contracted *flexor tendons,* inferior check *ligament desmotomy* or *tendon* lengthening may be necessary,[6] while angular limb deformities are treated by *metaphyseal* stapling or screw wire tension bands.[18]

A *legume—roughage ration*, without adequate amounts of a properly formulated *concentrate* mix, results in a phosphorus deficiency in the growing horse. Legumes contain about one half as much phosphorus as needed for growth (Table 8–1). Phosphorus deficiency may result in *epiphysitis*, angular limb deformities, or both.

A *grass—roughage ration*, without adequate amounts of a properly formulated *concentrate* mix, is deficient in all of the major *nutrients* needed for growth, *lysine, protein,* calcium, phosphorus, and *energy* (Table 8–1). Since it is deficient in all of these nutrients, growth rate is greatly reduced and, as a result, *epiphysitis* and contracted *flexor tendons* are not likely to occur. However, if the ration is not corrected, mature body size is reduced.

To prevent or treat these conditions, one must feed the correct amount (Table 8–3) of a properly formulated concentrate mix (Appendix Table 6).

Chapter 10
General Feeding and Management Practices

FEEDING

Good quality water and salt (preferably *trace-mineralized** salt) should always be available for all horses. Although mature, idle, nonlactating horses may be fed once daily, it is generally better to feed horses at least twice a day.

When and How to Feed

It is best to feed at shoulder level and on a regular schedule (Fig. 10–1). When the hay rack is too high, dust inhalation and respiratory problems increase; while feeding on the ground may result in fecal contamination and increased parasitism unless enclosed feeders are used. Hand feeding should be avoided, as it encourages nipping.

Many horses are fed in groups (Fig. 10–2). In this situation, horses low in the order of dominance in the group may be repeatedly driven away from feed and consequently may not receive an adequate amount. It is beneficial to place the feed in a large circle or in several areas in the pen, and to put partitions between horses. However, after dominant horses finish their feed, they will often drive subordinate horses from any remaining feed, so it may be necessary to separate particularly aggressive or subordinate horses for feeding (Fig. 10–3).

Always change feeds (type of hay, pasture, or *concentrate*, or the amount of each) gradually over several days. Whenever concentrate intake is increased, the increase should be made at a gradual rate. Usually, increasing concentrates at the rate of ½ lb (¼ kg) daily until the desired level is reached is the best approach. Increasing the amount of *grain* too rapidly may cause *founder* and *colic*. If the horse

---

*Words in italics are defined in the glossary.

154

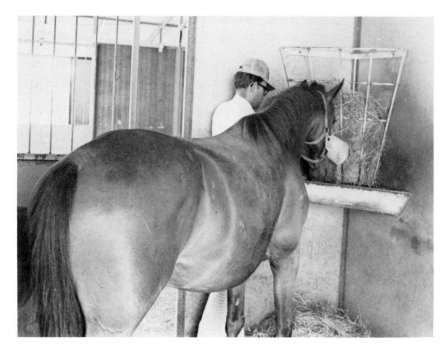

Fig. 10–1. Feeding both hay and grain at shoulder level or lower is preferred. Placing hay too high increases dust inhalation during eating, which increases coughing and heaves. Feeding on the ground increases fecal contamination of the feed, which increases ingestion of parasite eggs that are passed in the feces. This increases intestinal parasitism, or worms, in the horse.

has been receiving hay, give it all it wants immediately before turning the horse out on pasture, particularly if the pasture *forage* is lush and green. Increase the hours on pasture over several days if possible. Most horse prefer the feed they are accustomed to, so when the type of feed is changed, feed intake is reduced unless the new feed is much more palatable than the old.

Observe horses closely as they eat. Sudden changes in appetite are indicators of something wrong with the feed or with the horse.

Don't feed dusty or moldy feeds; they may cause *chronic* coughing, *heaves,* and *bleeders,* and may contain *mycotoxins* which can cause abortion and death. Moldy feeds may also be unpalatable. If poor quality feeds must be fed, feed a lot of them, so that the horse has the opportunity to pick out the best portions. If small amounts are fed, the horse is forced to eat the bad portions of the feed in an attempt to meet its nutritonal needs.

Most people feed by volume, i.e., by coffee cans, quarts, or gallons of *grain* and flakes of hay. *Cereal grains,* other *concentrates,* and hay may vary widely in density. Therefore, the weight of feed varies when

Fig. 10–2(A,B,C,D,E,F). Three methods of feeding hay and grain to horses in corrals or pastures. Tire hay feeders (A) may be made by removing most of the sides of all except the bottom side of the bottom tire and fastening the tires together. These work well, and generally result in the loss of less hay than wooden feeders (B). The greatest loss of hay occurs when it is fed on the ground (C); in addition, fecal contamination of the hay, resulting in increased intestinal parasites or worms, is also likely when it is fed on the ground. For these reasons, feeding on the ground is not recommended. Grain may be fed in wooden feeders (B), feeding bags (D), or in individual feeding pans or boxes (E). A feeding pan may be made from a single tire with part of the top side removed and the bottom fastened to a piece of wood (not shown). To prevent injury, metal feeding pans are not recommended. Feeding bags should be removed as soon as the horse finishes eating the grain. Feed bags are one of the best ways to feed grain to horses in a group (F). Each horse is assured of receiving the correct amount and type of grain mix.

Fig. 10–3. Effect of dominance when horses are fed as a group. The mare on the right is driving the less dominant mare away from the hay being fed. Less dominant horses may be repeatedly driven from feed, so that after a time they may make little effort to obtain their share, and become thin and in poor condition. To help prevent this when group feeding is necessary, feed in a large circle and ensure that there is a feeder for each horse. It may be necessary to group horses according to dominance, or to remove very dominant or submissive horses from the group to prevent this.

it is fed by volume, e.g., 1 qt, or a 1-lb coffee canful of oats or linseed meal, weighs 1 lb, whereas, 1 qt of bran weighs 0.5 to 0.6 lbs, and 1 qt of the other cereal grains or soybean meal weighs 1.5 to 1.9 lbs. A square bale of hay may weigh anywhere from 35 to 120 lbs, although it usually weighs 60 to 80 lbs, and a flake of hay will therefore generally weigh 5 to 10 lbs. This wide range in the weight of a bale of hay dramatically emphasizes the value of buying hay by weight rather than by the bale. The amount of feed needed by the horse can be much more closely estimated from the weight, rather than the volume of the feed. To determine the correct volume of feed, you should weigh several bales of hay to find an average weight, and weigh the amount of concentrate that your feed containers hold (Fig. 10–4).

ROUGHAGE

Provide all horses with a minimum of 0.75 to 1.0 lb of *roughage* per 100 lbs of body weight per day (0.75 to 1 kg/100 kg/day) in the form of good quality hay or pasture. Horses given access to pasture for as long as 4 hours daily will consume a sufficient amount of roughage to significantly reduce hay needs. Roughage adds bulk to the diet and tends to dilute *concentrate* feeds, thus preventing rapid fermentation

in the gut which may cause *colic* or *founder*. If the horse is to be used that day, feed it most of its hay *ration* at night, when the horse has time to eat it leisurely.

Lawn clippings may be fed when they are green and fresh, providing the grass has not been sprayed or treated with poisons of any type. However, clippings dry out quickly after cutting. Dry clippings are very dusty, and should not be fed unless they are pressed into pellets.

CONCENTRATES

Feed *concentrates* not less than twice daily. If the amount of concentrates fed at one time is more than 8 lbs (3.5 kgs), increase the frequency of feeding to 3 times per day at approximately 8-hour intervals, but maintain the same total daily intake.

OIL

Adding two oz (57 ml) of a vegetable oil or cooking oil, twice daily to the *ration* may put a shine on the horse's coat, and hasten shedding in the spring (Fig. 10–5). The type of vegetable oil used doesn't matter, since all are high in unsaturated fatty acids, which cause this effect. In contrast to vegetable oils, animal *fats* are low in unsaturated fatty acids. It may take 6 to 7 weeks before the full effect of adding oil to the ration is evident. Daily grooming, however, does more to hasten shedding and to put a *bloom* on a horse's coat. See the discussion on Feeding for Strenuous Physical Exertion in Chapter 5 for recommendations and effects of feeding a horse oil for this purpose.

Supplements

There are many *supplements* and conditioners on the market. Some have nutritional value, but are generally expensive for the *nutrients* provided. Many conditioners contain high amounts of unsaturated *fats* that may improve skin and coat condition (commonly called *bloom*), but cooking *oil* has a similar effect, and is much cheaper. Other

←———

Fig. 10–4(A,B,C,D). Buy feed and provide it to the horse according to weight, not volume. To do this, weigh several of the bales of hay being fed, and weigh the grain mix in the container used for feeding. The hay may be weighed by hanging it on the type of scale shown in A, or by standing on a bathroom scale and subtracting the difference with and without the bale (B). Any type of container may be used, such as a coffee can or grain scoop made by cutting the bottom out of a plastic jug (C,D).

Fig. 10–5. Adding 1 to 2 ounces of cooking oil to the grain at each feeding will help give the horse a shiny hair coat.

supplements may be worthwhile under certain conditions, such as stress, sickness, heavy lactation, or hard work, or when a particularly poor quality hay is being fed. Most commercial *vitamin-mineral* supplements contain many high-priced nutrients that are of absolutely no benefit if the horse is being fed average to good quality feeds. These supplements include substances such as vitamins D, E, K, C (ascorbic acid), thiamine (B_1), riboflavin (B_2), pangamic acid (B_{15}), pyridoxine (B_6), pantothenate, biotin, cyanocobalamin (B_{12}), folic acid, choline, iodine, zinc, cobalt, copper, sulfur, manganese, magnesium, potassium, chromium, molybdenum, para-aminobenzoic acid, inositol, methionine, lecithin, and iron. Some of these substances are not known to be needed by the horse, while others, although needed, are provided in ample quantities in nearly all commonly available *rations* (see sections in Chapter 1 on minerals and vitamins for a more complete discussion of these nutrients). Beware of feed supplements with unsubstantiated claims, such as testimonials, and studies with no controls, or nonsupplemented animals, fed the same ration without the supplement as a comparison.

If a nutritional problem is suspected, feeds should be analyzed for the *nutrients* of concern. The amount present in the *ration* should be compared with the amount needed, or that is toxic, and the ration corrected accordingly. Most commercial *supplements,* if not fed in excess of the manufacturer's recommendations, are unlikely to be detrimental to the horse. However, if the amount recommended is exceeded, or several supplements are being given, *vitamin* and *mineral* toxicities may occur. Commercially prepared *concentrate* mixes may contain added vitamins and minerals, which in conjunction with additional supplements added to the horse's ration, may result in toxicities. With the exception of *protein*, calcium, and phosphorus during pregnancy, lactation, and growth, if the horse is receiving average to good quality feeds, supplements of all types are of little or no benefit and add unnecessarily to the cost of the ration. Selenium would be the exception in areas where selenium deficiency is a problem.

Rumensin

Rumensin (monensin sodium) is a feed additive widely used to increase feed efficiency and rate of gain of cattle, and as a coccidiostat for poultry. It is present in many *grain* mixes intended for cattle and poultry, and may inadvertently be present in those fed to horses. Rumensin is highly toxic to the horse. At low concentrations in the horse's *ration* (30 ppm or less) its only effect may be to decrease feed intake and cause an uneasiness resembling mild *colic*. At higher levels (minimum lethal intake is 1 mg/kg body weight, and 2.5 mg/kg is fatal 50% of the time), the signs are posterior weakness, staggering, profuse sweating (pools of sweat may be produced for several days), symptoms resembling *azoturia*, and eventually inability to rise after going down. Other effects are blindness, head pressing, and an absence of muscle twitching (as often occurs from poisoning with heavy metals or chlorinated hydrocarbons). Death may occur within 12 to 24 hours after the onset of symptoms. The horse remains alert, and may thrash prior to death.

Blood concentrations of *urea* nitrogen, total bilirubin, and SGOT and alkaline phosphatase (enzymes released from damaged tissues) are elevated in affected horses. Necropsies usually show hepatic lipidosis, nephrosis, and evidence of *acute* cardiovascular collapse, with hemorrhages in the heart muscle, myocardial degeneration, and excessive accumulation of transudate in the abdominal cavity. Rumensin toxicity must be confirmed by laboratory analysis of the horse's feed or stomach contents. There is no known treatment other than removing the feed containing rumensin, and giving drugs and *mineral oil* to speed the passage of already ingested feed. Horses that survive may have an audibly abnormal heart rhythm, and increased heart and respiratory rate. They may *stock up* and can die during physical exertion as long as weeks to years after the poisoning. To prevent this, complete stall rest for at least several months is recommended.

Hair and Plasma Nutrient Analysis

Hair analysis is of little benefit in assessing an animal's nutritional status or the adequacy of the *ration* in most *nutrients*. This includes all *vitamins*, calcium, phosphorus, magnesium, sodium, potassium, iron, copper, and manganese.[22,30,40,76] Dietary levels of calcium and phosphorus in ponies do not affect the amount of these *minerals* deposited in the hair.[76] Various levels of iron in the diet of swine have produced no change in hair iron content.[30] Normal levels of copper in the hair vary too widely to make analysis of much use.[22] In horses, the content in the hair of all the minerals noted, plus zinc, varies depending on coat colors and different times of the year.[40] In addition, it is difficult to ensure an accurate analysis. Any dirt on the hair greatly alters the values obtained, as does excessive cleansing of the hair, which may leach out some of the minerals. The selenium and zinc content of the hair, however, may be decreased when these minerals are deficient in the *ration*,[4,42] and selenium may be increased when there is an excess in the ration (see Chapter 1). A decrease in hair root diameter and volume is indicative of a protein deficiency.[26,77]

Although changes in the concentration of some *nutrients* in the *plasma* (see Chapter 1 on zinc, copper and selenium), and *vitamin A* (of less than 20 μg per 100 ml) may be suggestive of a deficiency or excess of that nutrient in the *ration*, plasma, like hair analysis, is of little value in assessing the nutritional status of the animal or the adequacy of the ration for most nutrients. **The most accurate and reliable means of determining the nutritional status of the animal, or the adequacy of the ration is to determine the amount of nutrients present in the ration and to compare this amount to that required by the animal** (see Chapter 3).

MANAGEMENT

A number of miscellaneous management practices are discussed in this section.

Introducing New Horses

A physical barrier should be placed between strange horses when they are introduced. Ideally, horses should be put in adjacent paddocks so they can see, hear, and smell each other, but cannot kick each other. Although they can bite across the fence, one horse can easily escape the other. The dominant-subordinate relationship can be established between them with minimal chance of injury. Later, they can be placed in the same paddock. Some aggression may still occur, however, when they are placed together; therefore, avoid putting horses together in a crowded environment. Crowding increases aggressiveness, and since the loser cannot escape, it may be kicked repeatedly by the winner.

Preventing Stable Vices

Providing adequate and regular exercise is important in preventing stable vices such as stall kicking, *cribbing,* and *wood chewing.* Exercise and companionship to prevent boredom are the most beneficial factors in preventing these vices. Fat and a lack of exercise are two of the horse's (also affluent humans') worst enemies.

Effects of Exertion and Stress

Always warm up the horse slowly. Physical exertion without warming up the horse may cause *exertion myopathy,* also referred to as *azoturia, Monday-morning sickness,* and incorrectly, *tying-up syndrome.* In exertion myopathy the muscles become hard and painful, resulting in a stiff, stilted gait and reluctance to move. If this occurs, **the horse should not be moved,** not even a single step, and veterinary care should be obtained immediately. This condition occurs within a few minutes after physical activity has begun. Tying-up, which produces the same signs, occurs after prolonged physical activity, and is due to muscle *energy* depletion. Another condition called stress *tetany,* which also produces the same signs, may also occur after prolonged physical activity. Stress tetany occurs as a result of excessive loss of calcium, and occasionally magnesium, from the body, resulting in a decrease in their concentrations in the blood (Table 5–3). To treat the condition, a calcium-containing solution must be injected into the *vein.* This should be done very cautiously, while listening to the horse's heart, and only by a veterinarian. Giving calcium too fast or giving too much into the vein will cause sudden death.

Care of the Teeth

Check the teeth annually for points and float them when necessary. The horse's upper molars extend about one-half the width of a tooth outside the lower molars. The teeth wear at contact with each other, creating sharp points on the outside of the upper molars and on the inside of the lower molars. These points lacerate the cheeks and tongue, and make the horse's mouth sore. This causes slobbering of *grain,* tilting of the head when chewing, and results in poor physical condition. To prevent this, the teeth should be checked at least

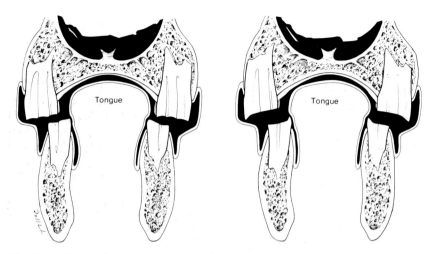

Fig. 10–6. Floating the teeth. The horse's upper molars, as shown here, extend about one-half a tooth width outside the lower molars. The teeth wear down where they come in contact with each other, progressing from the shape shown on the left to that shown on the right. The sharp points on the teeth shown on the right lacerate the cheeks and tongue, making the horse's mouth sore. A rasp is used to file these points off. This is called floating the teeth.

annually, and the points filed down. This is called *floating the teeth* (Fig. 10–6).

Care of the Hoofs

The hoofs should be trimmed every six weeks if the animal is shod; if not, the frequency of trimming depends on hoof wear. Horses should be shod for the following reasons: (a) to prevent excessive hoof wear when the animal is used on hard or rocky surfaces; (b) to correct faulty hoof structure or growth; (c) to aid in gripping the track or arena; and (d) to complement or correct the gait. If these factors are not present, the horse does not need to be shod. Many horses are shod that don't need to be. (See Chapters 7, 8, and 9 of Supplemental Reading Recommendation 5 for details on hoof care and shoeing.)

It is claimed that a number of nutritional factors influence hoof growth. These include various *vitamin* and *mineral supplements* and dietary additives. However, none, including gelatin, have been shown, in controlled studies, to have an effect on hoof growth if the animal is receiving a good *ration*. In addition, it is reported that commercial hoof dressings do not have a beneficial effect on hoof elasticity or growth.[5] If the *protein* content of the ration is below the animal's requirements, hoof growth may be slowed. Correcting this deficiency will increase the rate of hoof growth. Giving protein above the animal's requirements, however, will not increase the rate of hoof growth above that which occurs when the ration just meets the horse's needs. Under optimum conditions, hoof growth rate of nursing foals is approximately 0.6 inches (1.5 cm)/month; of yearlings, 0.5 inches (1.2 cm)/month; and of mature horses, 0.35 inches (0.9 cm)/month. Injury to or fever in the sensitive structures increases the rate of hoof growth.

Proper hoof moisture content is important in the maintenance of good hoofs. Factors that interfere with hoof moisture content, and as a result may cause hoof cracking and splitting, include stabling in sand lots or in deep manure or urine, excessive rasping of the sides of the hoof wall, turpentine, and most commercial hoof dressings. Factors that encourage ideal hoof moisture content include water, which horses can periodically, but not constantly, stand in (such as results from purposely overflowing the water trough), washing the hoofs each time they are regularly picked out, in some cases applying hoof dressing after moisture is added to the hoof, and daily packing saturated clay in the bottom of the hoof.

Bedding

It is important to maintain a dry, clean stall to prevent *thrush, grease heel* (Fig. 10–7), *ringworm* (Fig. 10–8), and poor skin and hair coat condition. Good drainage, clean, dry bedding, and frequent cleaning

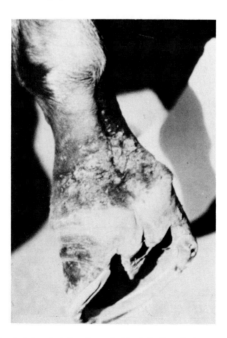

Fig. 10–7. Grease heel. A bacterial or fungal skin infection that may occur as a result of the horse standing in moist, dirty bedding.

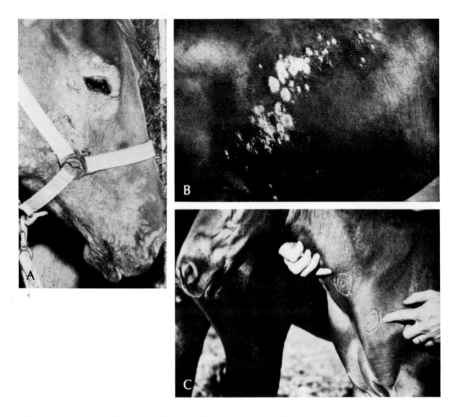

Fig. 10–8(A,B,C). Ringworm. Three different-appearing skin lesions, all of which are due to ringworm. Thiabendazole or iodine solution rubbed on the lesion with a toothbrush is generally effective. Be careful not to use brushes, combs, or tack from an affected animal on another animal, as the infection is readily spread in this manner. Frequently, it is necessary to bathe the entire horse to keep the disease from spreading. The animal should be treated daily for a week and then every other day for another week.

of the stall are necessary. Peat, wood shavings or *straw* are preferred for bedding. Straw is the preferred bedding for foaling mares. Shavings are more likely to be drawn into the vagina during foaling, and are more abrasive. Wheat, barley, and rye straw are best. Oat straw is less absorbent and more palatable, and rice straw is too stiff and wiry.

If *chronic* coughing, *heaves*, or bleeding from the nostrils are a problem, *straw* should not be used. To prevent coughing and heaves, it is best not to keep horses in the same building in which hay or straw is stored, particularly if it is kept in a loft. If the horse must be housed, let it out of the stall during the day. If at all possible, leave the top half of the box stall door open to the outside.

If wood shavings, chips, or sawdust are used, extreme caution should be taken to ensure that they do not contain black walnut. Skin contact with black walnut shavings is enough to cause *acute founder.* They do not need to be ingested, although the toxicity is worsened if they are eaten. Twelve to 24 hours after contact with fresh black walnut shavings, the horse may develop an acute, severe founder, and may die. In some cases, there may be milder signs of founder, increased temperature, and edema, or *stocking up,* of the knees and hocks.

Black walnut is a rich, dark-colored hardwood used for furniture and gunbarrel stock. Its shavings may range in color from purplish-black to coffee-brown and usually have a distinct sweet or acrid odor when fresh, which is when they are the most toxic. In contrast, pine and fir shavings are light-colored. Although not all dark-colored shavings are black walnut, it is safest not to use any wood shavings that contain any dark wood unless you know for certain that it is not black walnut. To determine whether shavings are black walnut, they may be sent to USDA Forest Products Laboratory Center for Wood Anatomy Research, Madison, Wisconsin 53705.

Hay should not be used for bedding because many horses will eat it, and because it is not very absorbent. Peanut hulls are absorbent, but some horses will eat them and they may attract rodents. Sugar cane residue is difficult to handle, and may be eaten. Corn *stover* is fairly absorbent, but should be closely inspected for mold, which may be toxic to the horse. Recycled newspapers, which are shredded and repulped to a cotton-like texture, make excellent bedding. They decrease stable odors, are highly absorbent, dry quickly, and are fluffy and comfortable for the horse to lie on.

Vaccination and Worming

Institute and maintain a good vaccination and worming program. This program will vary depending upon a multitude of factors including (1) disease and parasite prevalence in your area, (2) degree of confinement, (3) number of horses, (4) what the horses are used for, and (5) frequency of contact with other horses. Because of these many variables, vaccination and worming programs should be set up by your local veterinarian.

VACCINATION

A minimum vaccination program includes *tetanus* toxoid and *sleeping sickness* (equine encephalomyelitis) vaccines to all horses yearly. Depending on the incidence and potential exposure on each farm, equine *influenza* (flu), *rhinopneumonitis,* and strep equi *(strangles,* or *distemper)* vaccines may be indicated. Descriptions, pictures of horses affected by these diseases, and treatment are given in the glossary. An immunization program and worming program (as described below) for the foal should begin at 6 to 12 weeks of age. At the first vaccination, the foal, or any horse not previously vaccinated, may be given tetanus toxoid, sleeping sickness (encephalomyelitis), influenza and rhinopneumonitis vaccines and the vaccines repeated 1 to 2 months later. Strep equi (distemper or strangles) vaccine may also be given. It is not generally needed by horses over 2 years of age since they develop an immunity to the disease with or without prior vaccination. However, even in younger horses, frequently strep equi vaccine is not used because the vaccine is poorly effective and pain and swelling often occur at the injection site. Annual booster vaccinations for sleeping sickness and tetanus (using the toxoid not the antitoxin) are recommended for all horses.

To prevent respiratory disease and possibly a permanent decrease in respiratory function that may impair future athletic performance, it may be necessary to vaccinate for rhinopneumonitis and influenza every 2 months. This is recommended particularly for stables of horses used for athletic competition and in farms where the population changes frequently. On farms with a closed population, fewer vaccinations may suffice. Vaccination may be timed to coincide with the end of the race meet or show, so that horses have a break in performing after vaccination. Occasionally, fever, depression, and pain and swelling at the injection site may occur following influenza vaccination. Failure of a vaccination program against respiratory diseases is most commonly due to one or more of the following factors:

(1) vaccination for influenza and rhinopneumonitis less frequently than every 2 to 3 months,

(2) other viruses not present in the vaccine may be responsible, particularly rhinoviruses 1 and 2, adenovirus and arteritis virus, and

(3) incorrect handling of the vaccine, such as failure to keep it refrigerated, and use of outdated vaccines.

Pregnant mares should be vaccinated for rhinopneumonitis each year during the fifth, seventh, and ninth months of gestation. In addition, one month before foaling, they should be vaccinated for sleeping sickness and tetanus (using the toxoid not the antitoxin). The rhinopneumonitis killed-virus vaccine,* not the modified live-virus vaccine,† should be used for pregnant mares. The modified live-virus vaccine† may not, in some areas, give protection against abortion, at least.

---

* Pneumabort K (Fort Dodge Laboratories, Fort Dodge, IA 50501)
† Rhinomune (Norden Laboratories, Lincoln, NE 68521)

PREVENTION AND TREATMENT OF INTESTINAL PARASITES

Several universally applicable measures that should be practiced for intestinal parasite *(worms)* control are as follows:

(a) Fecal ingestion should be prevented, since many intestinal parasites are transmitted in this manner. Manure should be removed from stables, corrals, and small pasture lots frequently. Manure should be composted before spreading on pastures, or it should be spread on cropland or other ungrazed areas. Overstocking and overgrazing pastures should be avoided. Weanlings and yearlings should be kept separate from older horses. Horses should not be fed on the ground, and clean, non-fecally contaminated water should be provided to them.

(b) All horses on the farm should be included on the same *deworming* program. Mature horses should be treated at the same time and with the same drug as young horses.

(c) Transient and newly added horses should be quarantined, or isolated, and *dewormed* before they are turned out with the other horses (Fig. 10–9).

(d). Laboratory examination of fecal samples for *worm* eggs, or oocysts, should be made annually to maintain surveillance on the effectiveness of the drugs being used and of the worming program (Figs. 10–10, 10–11).

(e) The four different classes of drugs available for *deworming* should be used on a rotational basis. This is important in preventing

Fig. 10–9. Isolation pen. Most infectious diseases have an incubation period of 7 to 14 days from the time the animal is infected until clinical signs appear. During this time, as well as after clinical signs are present, contagious diseases are readily spread to other animals. To prevent this, all horses new to a stable should be dewormed, vaccinated, and kept isolated from all other horses for several weeks. All stables should have isolation facilities for this purpose.

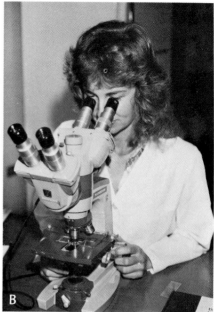

Fig. 10–10(A,B). A fecal flotation to determine the presence of intestinal worms. A small amount of feces is mixed in a magnesium sulfate solution. A microscope slide cover slip is placed on top of the tube, in contact with the solution (A). Eggs from intestinal worms, if present in the feces, float to the top of the solution and adhere to the cover slip. The cover slip is placed on a glass slide and viewed under a microscope (B).

the development of strains of parasites resistant to the drugs. Deworming products containing the same type of drug are called by different names by each manufacturer producing them. Therefore, rely on the generic name of the active drug listed on the label and not on the manufacturer's name for the product. The four different classes of drugs presently available are (1) the bendazoles, (2) the organophosphates, dichlorvos and trichlorfon, (3) the phenothiazine-piperazines, and (4) the carbamates. The bendazoles include mebendazole (Telmin), thiabendazole (Omnizole, Equizole, and Equi-Vet TZ), cambendazole (CamVet), fenbendazole (Panacur), oxfendazole (Benzelmin), and oxibendazole (Anthelcide). These are all highly effective against *strongyles* and *oxyuris*, and cambendazole and mebendazole are also highly effective against *ascarids*. They all have a high margin of safety, even in pregnant mares and foals.

The organophosphates, dichlorvos (Equigard and Equigel) and trichlorfon (Combot), are effective against *ascarids, pinworms (oxyuris)* and *bots (Gastrophilus)*. Dichlorvos is also effective against

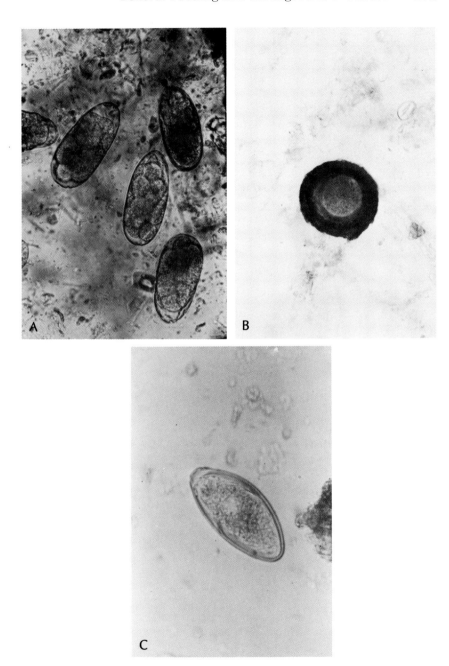

Fig. 10–11(A,B,C).   Eggs from intestinal worms are viewed under the microscope with 100 times magnification. Shown here are strongyles eggs (A), an ascarid (B), and a pinworm (oxyuris) egg (C).

strongyles. The organophosphates are potentially toxic and should not be given to foals under four months of age or to mares during the last three months of pregnancy. The phenothiazine-piperazine with carbon disulfide combination (Parvex Plus) is highly effective against strongyles, ascarids, pinworms, and bots. Piperazine alone is effective against ascarids, but is not effective against large strongyles or bots. Phenothiazine is effective against only strongyles and carbon disulfide is effective only against bots. The organophosphates and carbon disulfide may cause mild *colic* and a slight transient diarrhea following administration. The carbamates, such as pyrantel pamoate (Strongoid T) and pyrantel tartrate (Banminth and Strongoid), are effective against ascarids, strongyles and tapeworms (Anoplocephala). Pyrantel is well accepted when given by oral syringe, especially in foals. The bendazoles and pyrantel are preferred in sick or stressed horses, mares in late pregnancy, and in foals. Products consisting of combinations of these four products are also available, e.g., piperazine, trichlorfon, and phenothiazine. (Dyrex T.F.) or thiabendazole and piperazine (Equizole A).

(f) Labels should be read before a *deworming* medication is administered. All recommendations, precautions, and contraindications should be followed.

(g) *Deworming* medication may be added to the feed, given by stomach tube, or given as a paste which is put into the back of the horse's mouth with a syringe. One method is no more or less effective than another if all of the medication reaches the animal's stomach within a fairly short period of time. Thus, dewormers added to the feed are effective only if all of the dewormer is consumed within one hour or less. An effective dewormer that is injected into the muscle is presently being tested and may be available in the near future (Ivermectin, Merck and Co., Rahway, N.J.). It has been shown to be 97 to 100% effective against both the adult and larval stages of numerous parasites, including large and small strongyles, ascarids, pinworms, and bots.[102]

Deworming every two months with drugs effective against *strongyles, ascarids,* and *pinworms* is recommended for valuable horses in highly concentrated groups, such as many brood mare bands. In addition, every four months a drug effective against bots should be used. Less valuable, or less concentrated horses, should be treated every three to four months with drugs effective against strongyles, ascarids, pinworms, and bots. The minimal deworming program is twice a year, in June and December, using drugs effective against strongyles, ascarids, pinworms, and bots both times.

Foals are free of parasites at birth; however, they are susceptible, and infections may be acquired during the first one to two weeks of life. Regular treatment of mares is necessary to prevent this. The foal generally should be treated for both *ascarids* and *strongyles* beginning at six to eight weeks of age with repetitions every two months, or as is necessary for the other horses.

Parasites are the predisposing cause of 90% of all *colics* occurring in the horse. They damage the liver, lungs, blood vessels, intestines, and stomach, cause blood loss, and decrease digestive function. They may create impactions and perforations of the stomach and intestine, resulting in death. They may also cause *anemia,* diarrhea, weight loss, and reduced stamina or performance. Horses affected early in life may have their entire future health and performance impaired. Descriptions and pictures of each of these *gastrointestinal* parasites are given in the glossary.

References

1. Adams, O. R.: Lameness in Horses. Philadelphia, Lea & Febiger, 1974, pp. 198–200.
2. Anonymous, Hoard's Dairyman, April 25, 1978, p. 548.
3. Baker, H. J., and Lindsey, J. R.: Equine goiter due to excess dietary iodine. J. Am. Vet. Med. Assoc., 153:1618–1629, 1968.
4. Brossard, G. A., Carson, R. B., and Hidiroglow, M.: Influence of selenium on the selenium content of hair and on the incidence of nutritional muscular disease in beef cattle. Can. J. Anim. Sci., 45:197, 1965.
5. Butler, K. D.: Hoof care. Arab. Horse J., Sept., 1977, p. 136.
6. Coffman, J. R.: Epiphysitis in weanlings. Mod. Vet. Prac., 4:53–56, 1973.
7. Coffman, J. R.: Clearance ratios in the horse. Denver, Colo. Vet. Med. Assoc. Seminar, January 12, 1978.
8. Cook, W. R.: Upper respiratory disease. Proc. 40th Annual Conf. for Veterinarians, Colorado State University, January, 1979, pp. 168–173.
9. Crampton, W. W.: Rate of growth in draft colts. J. Agr. Hort., 26:172, 1923.
10. Dickson, W. B.: Endocrine glands. In Duke's Physiology of Domestic Animals. 8th Ed. Edited by M. J. Swenson. Ithaca, Comstock Pub. Assoc., 1970, pp. 1189–1252.
11. Drew, B., Barber, W. P., and Williams, D. G.: The effect of excess dietary iodine on pregnant mares and foals. Vet. Rec., 97:93–95, 1975.
12. Driscoll, J., Hintz, H. F., and Schryver, H. F.: Goiter in foals caused by excessive iodine. J. Am. Vet. Med. Assoc., 173:858–859, 1978.
13. Dunn, T. G.: Relationship of nutrition to reproductive performance and economics of beef production. Proc. Annual Meeting of Society for Theriogenology, Cheyenne, Wyoming, September 25, 1975, pp. 1–8.
14. Ekman, L.: Variation of some blood biochemical characteristics in cattle, horses and dogs, and causes of such variation. Ann. Rech. Vet., 7:125–128, 1976.
15. Eyre, P.: Equine pulmonary emphysema: a bronchopulmonary mould allergy. Vet. Rec., 91:134–140, 1972.
16. Filmer, J. F.: Enzootic marasmus of cattle and sheep. Aust. Vet. J., 9:163, 1933.
17. Flodin, L., and Tyznik, W. J.: The effect of two vs. four month weaning on the growth rate of foals. Proc. 5th Equine Nutrition and Physiology Symposium, 1977, p. 99.

175

18. Fretz, P. B., Turner, A. S., and Pharr, J.: Retrospective comparison of two surgical techniques for correction of angular deformities in foals. J. Am. Vet. Med. Asso ·., *172*:281–286, 1978.

19. Ginther, O. J.: Occurrence of anestrus, estrus, diestrus and ovulation over a 12-month period in mares. Am. J. Vet. Res., *35*:1173–1179, 1974.

20. Godbee, R. G.: Non-Protein Nitrogen Utilization by the Equine. Ph.D. Thesis, Dept. of Animal Science, Colorado State University, Fort Collins, July, 1978.

21. Goorich, R. D., Pamp, D. E., and Meiske, J. C.: Free choice minerals. Proc. Minnesota Nutrition Conf., Sept., 1977, pp. 171–177.

22. Gregor, J. L. M., et al.: Nutritional status of adolescent girls in regard to zinc, copper and iron. Am. J. Clin. Nutr., *31*:269, 1978.

23. Grondalen, T.: Osteochondrosis and arthrosis in pigs. III. A comparison of the incidence in young animals of the Norwegian Landrace and Yorkshire breeds. Acta Vet. Scand., *15*:43–52, 1973.

24. Grondalen, T., and Grondalen, J.: Osteochondrosis and arthrosis in pigs. IV. Effect of overloading on the distal epiphyseal plate of the ulna. Acta Vet. Scand., *15*:53–60, 1974.

25. Grondalen, T., and Vangen, O.: Osteochondrosis and arthrosis in pigs. V. A comparison of the incidence in three different lines of Norwegian Landrace breed. Acta Vet. Scand., *15*:61–79, 1974.

26. Haaland, G. L., Matsushima, J. K., Nockles, C. F., and Johnson, D. E.: Bovine hair as an indicator of calorie-protein status. J. Anim. Sci., *45*:826, 1977.

27. Hamblton, P.: Effect of Fat in the Diets of Horses for Endurance Activity. MS Thesis, Department of Animal Science, Colorado State University, Fort Collins, September, 1978.

28. Harmon, B. G.: Bioavailability of phosphorus in feed ingredients for swine. Feedstuffs, June 20, 1977, pp. 16–17.

29. Harrington, D. D., Walsh, J., and White, V.: Clinical and pathological findings in horses fed zinc deficient diets. Proc. 3rd Equine Nutrition and Physiology Symposium, 1973, p. 51.

30. Hedges, J. D., and Kornegay, E. T.: Interrelationship of dietary copper and iron as measured by blood parameters, tissue-stores, and feedlot performance of swine. J. Anim. Sci., *37*:1151, 1973.

31. Hedhammar, A., et al.: Overnutrition and skeletal disease. An experimental study in growing Great Dane dogs. Cornell Vet., *64*(Suppl. 5): 1–160, 1974.

32. Hintz, H. F.: Equine Nutrition seminar presented at Colorado State University Annual Conf. for Veterinarians, January, 1977, Fort Collins, Colorado.

33. Hintz, H. F., et al.: Feedstuffs, March 20, 1978, p. 27.

34. Hintz, H. F., and Schryver, H. F.: Nutrition and bone development in horses. J. Am. Vet. Med. Assoc., *168*:36–44, 1976.

35. Houpt, K. A.: Equine maternal behavior and its aberrations. Equine Pract., *1*:7–20, 1979.

36. Hoversland, A. S., Harris, R., Krum, L., and Murphy, S.: Growth and Quarter Horse foal. I. Maturation of the serum biochemical profile from birth to twenty-four weeks. Proc. 5th Equine Nutrition and Physiology Symposium, 1977, pp. 100–101.

37. Isaacks, R. E., et al.: Restricted feeding of broiler type replacement stock. Poult. Sci., *39*:339–346, 1960.

38. Jordon, R. M., Myers, V. S., Yoho, B., and Spurrell, F.: A note on calcium and phosphorus levels fed ponies during growth and reproduction. Proc. 3rd Equine Nutrition Symposium, 1973, pp. 55–70.
39. Kasstrom, H.: Nutrition, weight gain and development of hip dysplasia. An experimental investigation in growing dogs with special reference to the effect of feeding intensity. Acta Radiol. (Suppl.), *334*:135–179, 1975.
40. Kossila, V., Virtanen, E., Hakatie, H., and Luoma, E.: Calcium, magnesium, sodium, potassium, zinc, iron, copper and manganese in the hair and mane of horses. J. Scientific Agric. Soc. Finland, *44*:207–216, 1972.
41. Krook, L., and Lowe, J. E.: Nutritional secondary hyperparathyroidism in the horse. Pathol. Vet., *1*:44–87, 1964.
42. Lewis, P. K., Jr., Hoekstra, W. G., and Grummer, R. H.: Restricted calcium feeding versus zinc supplementation for the control of parakeratosis in swine. J. Anim. Sci., *16*:578, 1957.
43. Linkswiler, H. M., Hoyce, C. L., and Anand, C. R.: Calcium retention of young adult males as affected by level of protein and of calcium intake. Trans. NY Acad. Sci., *36*:33–340, 1974.
44. Margen, S., Chu, J. Y., Kaufman, N. A., and Calloway, D. H.: Studies in calcium metabolism. I. The calciuretic effect of dietary protein. Am. J. Clin. Nutr., *26*:584–589, 1974.
45. Mayhew, I. G., et al.: Spinal cord disease in the horse. Cornell Vet., *68*(Suppl. 6):1–175, 1978.
46. McIlwraith, C. W., and Fessler, J. F.: Evaluation of inferior check ligament desmotomy for treatment of acquired flexor tendon contracture in the horse. J. Am. Vet. Med. Assoc., *172*:293–298, 1978.
47. Milne, D. W., Muir, W. W., and Skarda, R. T.: Effects of training on heart rate, cardiac output and lactic acid in Standardbred horses using a standardized exercise test. Equine Med. Surg., *1*:131–135, 1977.
48. Mosier, J. E.: Nutritional recommendations for gestation and lactation in the dog. Vet. Clin. North Am., *7*:683–692, 1977.
49. Nehrich, H., and Stewart, J. A.: The effects of prenatal protein restriction on the developing mouse cerebrum. J. Nutr., *108*:368–372, 1978.
50. Nutrient Requirements of Horses. 4th rev. ed. National Academy of Sciences, Washington, D.C., 1978.
51. Olsson, S. E.: Osteochondrosis—a growing problem to dog breeders. Gaines Progress, Summer, 1976, pp. 1–11.
52. Owen, J. M.: Abnormal flexion of the corono-pedal joint of "contracted tendons" in unweaned foals. Equine Vet. J., *7*:40–45, 1975.
53. Pearce, G. A.: The nutrition of racehorses: A review. Aust. Vet. J., *51*:14, 1975.
54. Persson, S.: Value of haemoglobin determination in the horse. Nord. Vet. Med., *21*:513–523, 1969.
55. Potter, G. D., and Huchton, J. D.: Growth of yearling horses fed different sources of protein with supplemental lysine. Proc. 4th Equine Nutrition and Physiology Symposium, 1975, pp. 19–21.
56. Reed, K. A., and Dunn, N. K.: Growth and development of the Arabian horse. Proc. 5th Equine Nutrition and Physiology Symposium, 1977, pp. 76–98.
57. Rooney, J. R., and New Castle, D. E.: Forelimb contracture in the young horse. Equine Med. Surg., *1*:350–351, 1977.
58. Rothman, E. E., and Sprock, H. A.: Horse pastures. Colorado State University Extension Bulletin No. 102, 1976.

59. Rumbaugh, G. E., Ardans, A. A., Gino, D., and Trommershausen-Smith, A.: Identification and treatment of colostrum-deficient foals. J. Am. Vet. Med. Assoc., *174*:273–276, 1979.
60. Saville, P. D., and Lieber, C. S.: Increases in skeletal calcium and femur cortex thickness produced by undernutrition. J. Nutr., *99*:141–144, 1969.
61. Schryver, H. F., Hintz, H. F., and Craig, P. H.: Phosphorus metabolism in ponies fed varying levels of phosphorus. J. Nutr., *101*:1257–1264, 1971.
62. Schryver, H. F., and Hintz, H. F.: Recent developments in equine nutrition. Anim. Nutr. and Health, *4*:6–10, 1975.
63. Schryver, H. F., Hintz, H. F., and Lowe, J. E.: Calcium and phosphorus in nutrition of the horse. Cornell Vet., *64*:491–515, 1974.
64. Schryver, H. F., and Hintz, H. F.: Feeding horses. Extension Information Bulletin 94, New York State College of Agriculture and Life Sciences, Ithaca, 1975.
65. Schryver, H. F.: Bending properties of cortical bone of the horse. Am. J. Vet. Res., *39*:25–28, 1978.
66. Schryver, H. F., Van Wie, S., Daniluk, P., and Hintz, H. F.: The voluntary intake of calcium by horses and ponies fed a calcium deficient diet. Equine Med. Surg., *2*:337–340, 1978.
67. Siassi, F., and Siassi, B.: Differential effects of protein-calorie restriction and subsequent repletion on neuronal and non-neuronal components of cerebral cortex in newborn rats. J. Nutr., *103*:1625–1633, 1973.
68. Slade, L. M., and Lewis, L. D.: Nutritional adaptation of horses for endurance performance. Proc. 4th Equine Nutrition and Physiology Symposium, 1975, pp. 14–15.
69. Smith, R. R., Rumsey, G. L., and Scott, M. L.: Heat increment associated with dietary protein, fat, carbohydrate and complete diets in salmonids: comparative energetic efficiency. J. Nutr., *108*:1025–1032, 1978.
70. Traver, D. S., et al.: Clearance ratios in the evaluation of electrolyte and mineral metabolism in horses. Proc. 5th Equine Nutrition and Physiology Symposium, 1977, pp. 15–23.
71. Traver, D. S., et al.: Renal metabolism of endogenous substances in the horse. Volumetric versus clearance ratio methods. Equine Med. Surg., *1*:378–382, 1977.
72. Ullrey, D. E., Struthers, R. D., Hendricks, D. G., and Brent, B. E.: Composition of the mare's milk. J. Anim. Sci., *25*:217–222, 1966.
73. Voss, J. L., and Pickett, B. W.: Effect of nutritional supplementation on pregnancy rate in nonlactating mares. J. Am. Vet. Med. Assoc., *165*:702–703, 1974.
74. Wilson, T. N., and Osbourn, D. F.: Compensatory growth after undernutrition in mammals and birds. Biol. Rev., *50*:324–373, 1960.
75. Wiltbank, J. N.: Effects of nutrition on reproductive performance of beef cattle. Am. Assoc. Bovine Pract., Proc. Bovine Nutrition Conf., Chicago, Sept. 12, 1975.
76. Wysocki, A. A., and Klett, R. H.: Hair as an indicator of the calcium and phosphorus status of ponies. J. Anim. Sci., *32*:74–78, 1971.
77. Zain, B. K., Haquani, A. H., Oureski, N., and el Nisa, I.: Studies on the significance of root hair protein and DNA in protein calorie malnutrition. Am. J. Clin. Nutr., *30*:1094, 1977.
78. Klendshof, C., Potter, G. D., Lichtenwalner, R. E., and Householder, D. D.: Nitrogen digestion in the small intestine of horses fed crimped or micronized sorghum grain or oats. Proc. Texas Ag. Conf., Texas A & M, April, 1980.

79. Nutrients and Toxic Substances in Water for Livestock and Poultry. Washington, D.C., National Academy of Sciences, 1974, pp. 1–93.
80. Maylin, G. A., Rubin, D. S., and Lein, D. H.: Selenium and vitamin E in horses. Cornell Vet., 70:272–289, 1980.
81. Basler, S. E., and Holtan, D. W.: Factors affecting selenium levels in horses throughout Oregon and influence of selenium levels on disease incidence. J. Science (Abstract), 1981.
82. Blackmore, D. J., Wilett, K., and Agness, D.: Selenium and gamma-glutamyl transferase activity in the serum of thoroughbreds. Res. Vet. Sci., 26:76–80, 1979.
83. Stowe, H. D.: Serum selenium and related parameters of normally and experimentally fed horses. J. Nutrition, 93:60–64, 1967.
84. Caple, I. W., et al.: Blood glutathione peroxidase activity in horses in relation to muscular dystrophy and selenium in nutrition. Aust. Vet. J., 54:57, 1978.
85. Messer, N.T.: Tibiotarsal effusion associated with chronic zinc intoxication in three horses. J. Am. Vet. Med. Assoc., 178:294–297, 1981.
86. Copper: Medical and Biologic Effects of Environmental Pollutants. Washington, D.C., National Academy of Sciences, 1977, pp. 1–115.
87. Underwood, E. J.: Trace Elements in Human and Animal Nutrition. 3rd Ed. New York, Academic Press, 1971, p. 82.
88. Smith, J. D., Jordan, R. M., and Nelson, M. L.: Tolerance of ponies to high levels of dietary copper. J. Anim. Sci., 41:1645, 1975.
89. Carberry, J. T.: Osteodysgenesis in a foal associated with copper deficiency. NZ Vet. J., 26:279, 1978.
90. Egan, D. A., and Murrin, M. P.: Copper responsive osteodysgenesis in a thoroughbred foal. Irish Vet. J., 27:61–62, 1973.
91. Stowe, H. D.: Effects of age and impending parturition upon serum copper of thoroughbred mares. J. Nutr., 95:179–183, 1968.
92. Wagenaar, G.: Iron dextran administration to horses. Tijdschr. Diergeneeskd, 100:562–563, 1975.
93. Donoghue, S., Kronfeld, D. S., Berkowitz, S. J., and Copp, R. L.: Vitamin A nutrition of the equine: growth, serum biochemistry and hematology. J. Nutr., 111:365–374, 1981.
94. Haeger, K.: Long-time treatment of intermittent claudication with vitamin E. Am. J. Clin. Nutr., 27:1179–1181, 1974.
95. Romanovschi, S., Popescu, F., Suciu, T., and Brumboiu, M.: Economic rations for young horses in training. Lucrarile Stiintifice al Institutului de Cercetari pentru Nutritie Animala, 1:367–376, 1972.
96. United States Department of Agriculture: The official United States Standards for Grain. Dec., 1975.
97. Heimann, E.: Selenium in Pregnant Mares Grazing Fescue Pastures. M.S. Thesis, Univ. of Missouri, Columbia, MO, 1980.
98. Garrett, Larry: Reproductive Problems in Pregnant Mares Grazing Fescue Pastures. M.S. Thesis, Univ. of Missouri, Columbia, MO, 1979.
99. Emiola, L., and O'Shea, J. P.: Effects of physical activity and nutrition on bone density measured by radiographic techniques. Nutr. Rep. Int., 17:669–681, 1978.
100. Bell, R. R., Tzeng, D. Y., and Draper, H. H.: Long-term effects of calcium, phosphorus and forced exercise on the bones of mature mice. J. Nutr., 110:1161–1168, 1980.
101. Glade, M. J., et al.: Growth Suppression and Osteochondrosis Dissecans in Weanlings Treated with Dexamethasone. Proc. American Assoc. Equine Pract. Ann. Meeting, 1979, pp. 361–365.

102. Bello, T. R., and Norflett, C. M.: Critical antiparasitic efficacy of Ivermectin against equine parasites. J. Equine Vet. Sci., 1:14–17, 1981.
103. Hintz, H. F.: Factors affecting the growth rate of horses. Horse Short Course Proc., Texas A & M Animal Agric. Conf., College Station, Texas 77843, 1979, pp. 1–5.

SUPPLEMENTAL READING

1. The following three excellent booklets on reproduction in the horse can be obtained from the Animal Reproduction Laboratory, Colorado State University, Fort Collins, CO 80523.
   (a) Voss, J. L., and Pickett, B. W.: Reproductive Management of the Brood Mare. 1976. 29 pp. $5.00.
   (b) Pickett, B. W., Squires, E. L., and Voss, J. L.: Normal and Abnormal Sexual Behavior of the Equine Male. 1981. 33 pp. $7.50.
   (c) Pickett, B. W., and Back, D. G.: Procedures for Preparation, Collection, Evaluation, and Insemination of Stallion Semen. 1973. 26 pp. $3.00.
   (d) Pickett, B. W., Voss, J. L., Squires, E. L., and Amann, R. P.: Management of the Stallion for Maximum Reproductive Efficiency. 1981. 84 pp. $15.50.
2. Nutritient Requirements of Horses. 4th rev. ed., 1978. National Academy of Sciences, Printing and Publishing Office, 2101 Constitution Avenue, Washington, D.C. 20418. $4.50.
3. Evans, J. W., Borton, A., Hintz, H. F., and Van Vleck, L. D.: The Horse, 1977, 766 pp., W. H. Freeman and Co., 660 Market St., San Francisco, CA 94104. Covers modern horse management and ownership, includes sections on nutrition, history, breeds, use of the horse, anatomy and physiology, selection, reproduction, genetics, health, management, horsemanship, foot care, fences and buildings. Recommended for the horseman.
4. Ensminger, M.E.: Horses and Horsemanship. 5th Ed., 1977, 546 pp. Interstate Publishers and Printers, Danville, IL 61832. Covers horse ownership, management, business aspects of horse production, health, disease prevention, and parasite control.
5. Adams, O. R.: Lameness in Horses. 3rd Ed, 1974, 566 pp. Lea & Febiger, 600 Washington Square, Philadelphia, PA 19106. The best book available on the relationship between conformation and lameness, soundness exam, diagnosis and treatment of lameness, and foot care, including trimming and shoeing.
6. Emery, L., et al.: Horseshoeing Theory and Hoof Care. Philadelphia, Lea & Febiger, 1977.

Glossary

A

Abscess: A localized collection of pus in the tissues of the body, often accompanied by swelling and inflammation, and generally caused by bacteria (Glossary Fig. 3).

Acute: Occurring over a short period of time; an acute disease is one which develops and progresses to death or recovery quite rapidly. Opposite of *chronic.* Often incorrectly used to indicate severity.

Agglutination: A phenomenon in which the cells distributed in fluid collect into clumps.

Air dry: A feed that has been allowed to dry in the air. As a result, it contains 8 to 12% moisture, depending on humidity.

Air dry basis: Indicates that the value expressed is the amount by weight present in an air dried feed or *ration.* Nearly all feeds for the horse, with the exception of green pasture *forage* and ensiled feeds, are air dried. Green growing pasture forage, *silage* and *haylage* generally contain about ⅔ water and ⅓ *dry matter.* See *dry matter basis* for conversion between *air dry, as fed* and dry matter basis.

Allergy: A condition of increased sensitivity to a specific *protein* (not an amount of protein) so that exposure to that protein causes an excessive response of the body which may be manifested in any one or more of the following effects: a rash, hives, respiratory difficulties, sneezing, nasal discharge, or diarrhea.

Amino acids: Nitrogen-containing compounds that are the "building blocks" from which *proteins* are formed. A protein consists of many amino acids hooked together by a peptide bond (⌇ as shown in the following).

$$
\text{H}-\overset{\displaystyle \text{H}}{\underset{\displaystyle \text{R}_1}{\text{N}}}-\overset{\displaystyle \text{H}}{\text{C}}-\overset{\displaystyle \text{O}}{\text{C}}-\!\!\!\sim\!\!\!\left(\overset{\displaystyle \text{H}}{\text{N}}-\overset{\displaystyle \text{H}}{\underset{\displaystyle \text{R}}{\text{C}}}-\overset{\displaystyle \text{O}}{\text{C}}\right)_{\!n}\!\!\!\sim\!\!\!-\overset{\displaystyle \text{H}}{\text{N}}-\overset{\displaystyle \text{H}}{\underset{\displaystyle \text{R}_2}{\text{C}}}-\overset{\displaystyle \text{O}}{\text{C}}-\text{OH}
$$

| N-terminal amino acid | many amino acids | carboxyl amino acid |

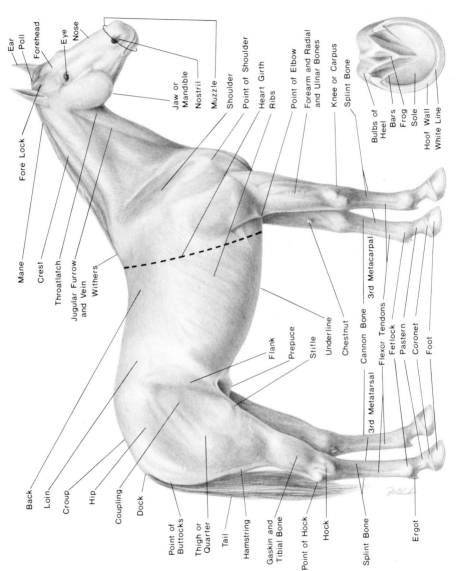

Glossary Fig. 1. Major external anatomic structures of the horse.

There are 22 different amino acids, i.e., 22 different R groups, as shown above. Different types of protein contain different amino acids. Twelve to 14 of these amino acids are produced in the body and do not need to be absorbed from the intestinal tract. These are called nonessential amino acids. The remaining 8 to 10 amino acids must be absorbed from the intestinal tract, and are called essential amino acids. The bacteria in the mature horse's large intestine produce all of the amino acids needed, so it doesn't matter which amino acids, or type of protein, is present in their *ration*. However, certain amino acids are needed in the ration for growth (see Chapter 1).

Amnionic sac: A translucent membranous sac in the *uterus* of the pregnant animal that contains fluid and the fetus.

Anemia: A decrease in *red blood cells* in the blood.

Anoplocephala: See *tapeworms*.

Antibiotics: Chemical substances produced by fungi that inhibit or destroy bacteria or other *infectious organisms*, and that are used primarily in the treatment of infectious diseases.

Antibodies: See *immunoglobulins*.

Antitoxin: Antibodies that counteract a toxin. These are given to an animal to protect it against that particular toxin. They do not stimulate the animal to produce its own antitoxins, and therefore provide protection against that toxin for only 1 to 3 weeks (see Vaccine).

Artery: A round, elastic, thick-walled tube that carries blood from the heart to all parts of the body. The blood in an artery is normally well saturated with oxygen it has picked up in its passage through the lungs. This gives it a bright red color, whereas the blood in a *vein* is dark red because the oxygen has diffused out of the blood into the tissue. Arteries are deeper from the body surface than are veins. The blood in an artery is under high pressure; therefore, if an artery is cut blood spurts out, whereas the blood in a vein is under very little pressure.

Ascarids (*roundworms*, Parascaris equorum): A white, round worm usually 6 to 9 inches (15 to 22 cm) in length. It affects primarily the young horse, although it may also infect older horses. Most horses past 3 to 4 years of age have developed an immunity to it. All horses should be treated for ascarids. The adult worms live in the intestine where they lay eggs which are passed in the manure (see Figs. 10–10 and 10–11). After the horse swallows the larvae that develop from these eggs the larvae migrate through the wall of the intestine, through the liver, and to the lungs where they are coughed up and reswallowed. This migration causes damage to the liver and lungs. Large numbers of ascarids in the intestine

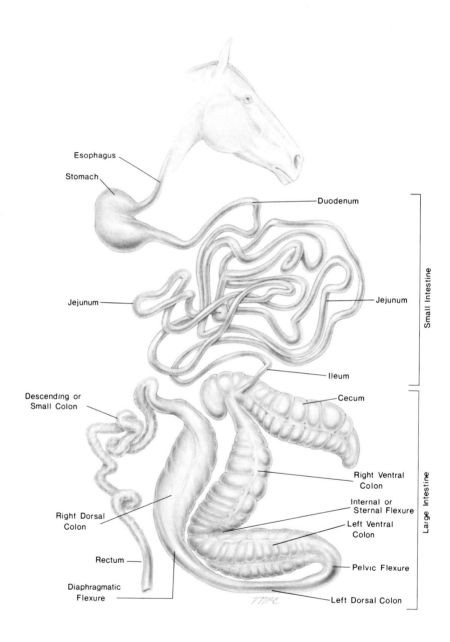

Esophagus

Stomach

Duodenum

Jejunum

Jejunum

Small Intestine

Ileum

Descending or
Small Colon

Cecum

Right Ventral
Colon

Internal or
Sternal Flexure

Right Dorsal
Colon

Left Ventral
Colon

Rectum

Pelvic Flexure

Diaphragmatic
Flexure

Left Dorsal Colon

Large Intestine

Glossary Fig. 2. Digestive tract of the horse. The stomach of the 1100-lb horse holds 2 to 3 gallons. Some protein digestion and partial breakdown of the feed occurs in the stomach. Liquids pass from the stomach rapidly, most leaving within 30 minutes after ingestion. Solid particles are retained for longer periods, to be broken down by the strong acid and digestive enzymes (pepsin) secreted by the stomach. The small intestine is about 70 ft (22 m) long, 3 to 4 inches (7 to 10 cm) in diameter, and holds 10 to 12 gal (40 to 50 L). Much of the fat and protein, and about one half of the soluble carbohydrate or nitrogen-free extract are digested in the small intestine. These and most of the vitamins and minerals are absorbed from the small intestine. Liquids pass through the small intestine rapidly, and reach the cecum 2 to 8 hours after ingestion. In another 5 hours, most of the liquid that reaches the cecum passes on into the colon. Passage of both liquids and particulate matter through the colon is slow and occurs over a period of about 50 hours. Nearly all the crude fiber or cellulose, and much (over 50%) of the soluble carbohydrate in feeds passes through the small intestine into the cecum. The cecum is about 4 ft (1.25 m) long and holds 7 to 8 gal (25 to 30 L). It contains bacteria which digest much of the fiber and about one half of the soluble carbohydrate (NFE) ingested. After digestion these nutrients are absorbed from the cecum and colon. Bacterial protein is also produced, digested, and absorbed from the cecum and colon. The large colon is 10 to 12 ft (3 to 3.7 m) long with an average diameter of 8 to 10 inches (20 to 25 cm) and holds 14 to 16 gal (50 to 60 L). It consists of four portions: (1) the right ventral colon, (2) the sternal flexure to the left ventral colon, (3) the pelvic flexure (where obstruction most commonly occurs) to the left dorsal colon, and (4) the diaphragmatic flexure to the right dorsal colon which connects to the small colon. The small colon is about 10 to 12 ft (3.5 m) long and 3 to 4 inches (7.5 to 10 cm) in diameter. When it enters the pelvic inlet it is called the rectum, which is about 1 ft (0.3 m) long and opens to the exterior at the anus. The large colon, small colon, and rectum make up the large intestine.

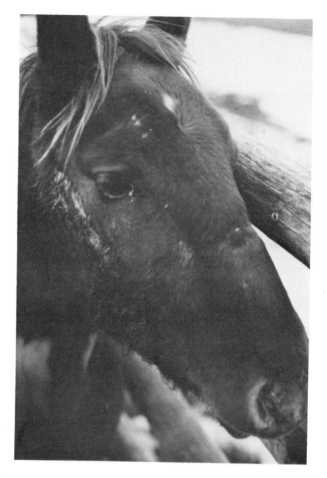

Glossary Fig. 3. An abscess to the left of the eye that has ruptured to exude a yellowish pus down the side of the face of a horse with strangles (distemper).

may cause impactions and *colic,* intestinal perforations, and death (Glossary Fig. 4). See discussion of intestinal parasites in Chapter 10 for treatment.

As fed or as is basis: Indicates that the value expressed is the amount present in the feed or *ration* in the form in which it is fed. The amount of *nutrients,* as given on the feed tag, are on an as fed basis. See *dry matter basis* for conversion between *air dry,* as fed, and *dry matter basis.*

Ash: See *minerals.*

Ataxia: Failure of muscular coordination, resulting in a stumbling, weaving, or drunken-type gait.

Azoturia (*exertion myopathy, Monday-morning sickness,* capture myopathy, paralytic hemoglobinuria): The excretion of nitroge-

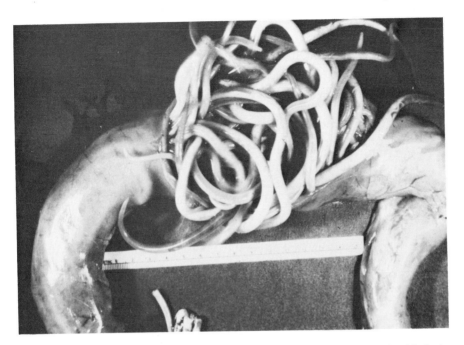

Glossary Fig. 4. Ascarids (Parascaris equorum) which became so numerous that they blocked and then ruptured the intestine, resulting in the death of the horse.

nous compounds in the urine. It is used to refer to a condition in the horse in which, within a few minutes of beginning physical activity, muscle spasms or *tetany* occur. Affected muscles become hard to the touch, muscle cells die, and muscle *proteins*, such as myoglobins, are lost and excreted in the urine. Sometimes incorrectly called the *tying-up syndrome* (see Table 5–3).

B

Balanced ration: A *ration* which provides the proper amounts and proportions of all the *nutrients* required by the animal.

Beta-carotene: A yellow plant pigment that, following ingestion and subsequent absorption, is converted by the animal to *vitamin* A.

Big-head (bran disease): Enlargement of the bones of the skull owing to replacement of their *minerals* with fibrous connective tissue. The jaw bone and the bones of the face are usually affected to the greatest extent, although all bones in the body are affected. The disease is brought on by excessive secretion of *parathyroid* hormone (parathormone), which may be caused by a dietary calcium deficiency or phosphorus excess, by parathyroid gland dysfunction, or by tumors (cancer) which may produce a substance which has the same effect as parathyroid hormone (see Fig. 1–5).

Bleeders: Horses that develop a bloody nose during exercise. The blood is coughed up from broken blood vessels in the lungs, resulting from *chronic bronchitis, heaves,* and scar tissue remaining from prior respiratory disease (see Chapter 4).

Blood counts: Generally refers to the practice of taking a blood sample and measuring its *hematocrit* and *hemoglobin* concentrations. The values obtained from the normal-appearing horse at rest are in most instances meaningless (see Chapter 5). If the hematocrit or hemoglobin concentrations are considered to be low, several *nutrients* such as iron, *vitamin* B, B₁₂, and copper needed for *red blood cell* synthesis are frequently given to the horse and a process called *"jugging"* may be performed. In most instances, these *nutrients* are not deficient and giving them or *"jugging"* the horse is of no benefit.

Bloom: When used in reference to an animal, it indicates a glossy hair coat, in reference to a plant it indicates the appearance of flowers.

Glossary Fig. 5. Botfly (Gastrophilus intestinalis) lays eggs called nits on the horse's hair (Glossary Fig. 9). After being swallowed, the eggs hatch, and bot larvae develop (Glossary Figs. 6 and 7).

Bone marrow: The soft material that fills the cavities of the bones. Red and white *blood cells* are produced by the bone marrow.

Bots (Gastrophilus species): Bot flies (Glossary Fig. 5) lay eggs on horses' hair in summer and early fall. The eggs, called *nits*, are about the size of a pin head and are yellow (Glossary Fig. 9). When the horse scratches itself with its teeth or licks itself, the eggs enter the mouth, and hatch. The larvae may invade the tissue of the mouth or be swallowed. Bot worms develop and attach to the wall of the stomach (Glossary Fig. 6). They cause inflammation of the stomach, and may cause perforations and death. The bot worms are passed in the manure in the spring, and are about ¼ by ¾ inch (½ cm by 2 cm) in size (Glossary Fig. 7). Bots can affect all horses of all ages. Bot flies are killed by the first hard freeze in the fall. See discussion of intestinal parasites in Chapter 10 for treatment.

Broken wind: See *heaves.*

Bronchial tubes (bronchi): Tubes, or air passages, formed by the division of the windpipe (trachea) at the point where it reaches the lungs.

Bronchitis: Inflammation of the *bronchial tubes.*

Glossary Fig. 6. Bots attached to the lining of the horse's stomach, which has been opened to expose them. They cause inflammation, and may cause perforations of the stomach, and death. After being passed in the manure in the spring, they pupate, and botflies develop (Glossary Fig. 5).

Glossary Fig. 7.   Bots. These develop in the horse's stomach from botfly eggs, or nits (Glossary Fig. 9).

Bronchiolytic agents: Drugs that help break up mucus in the *bronchial tubes.*
Buccal: Pertaining to the cheek or the mouth.
Bulk limited ration: A *ration* so low in *energy* density that the animal is unable to eat enough of it to meet its energy needs.

C

Calcitonin: A substance (hormone) produced by the thyroid gland in response to an increase in the blood calcium concentration. It inhibits calcium and phosphorus mobilization from the bone and decreases intestinal calcium absorption. In some animals it also increases calcium excretion in the urine. Its function is to prevent an increase in the blood calcium concentration.
calorie: A unit of *energy,* being the amount of heat required to raise 1 gram of water 1° C. This is known as the small calorie or standard calorie. This unit is not used in nutrition. The smallest unit of energy used in nutrition is the large calorie or kilocalorie. It is equal to 1000 small calories and may be written as 1 calorie, 1 Calorie of 1 kilocalorie.
Ca:P ratio (calcium:phosphorus ratio): The amount of calcium with respect to phosphorus present in the *ration,* with the amount of

phosphorus generally being given a value of one. For example, if a ration contained 0.70% calcium and 0.35% phosphorus, the Ca:P ratio of 0.70:0.35 would be converted to 2:1, indicating it contained twice as much Ca as P. This conversion is made by dividing both the amount of calcium and the amount of phosphorus in the ration by the amount of phosphorus.

Carbohydrates: Compounds composed of carbon, hydrogen, and oxygen whose major nutritional function is to supply *energy*. The most important carbohydrates in the horse's *diet* are starches, sugars, and *cellulose*.

Cardiac output: The amount of blood the heart pumps each minute.

Carnivores: See *herbivores*.

Carotene: See *beta-carotene*.

Carpus or carpal joint: The horse's knee (see Glossary Fig. 1).

Cation: An *inorganic* atom or molecule that when free in solution has a positive electrical charge. Opposite of anion which has a negative electrical charge. For example, when sodium chloride (common salt) dissolves in water the sodium and chloride molecules separate from each other leaving a positively charged sodium atom (cation) and a negatively charged chloride atom (anion). All charged atoms or molecules are called ions; e.g., sodium ion.

Cecum: A large compartment, or outpouching, of the intestinal tract of some animals, such as horses, rats, and rabbits. Ingested material passes from the mouth down the esophagus to the stomach, small intestine, cecum, large intestine or *colon*, rectum, and anus. The cecum contains many bacteria that digest much of the fibrous feeds that the animal ingests. It serves many of the same functions as the *rumen* in cattle and sheep. The appendix in humans is the vestige of a cecum (see Glossary Fig. 2).

Cellulose: A *carbohydrate* that forms the skeleton of most plants. Animals do not produce the enzymes necessary to digest it, whereas many bacteria do. Animals, such as horses and *ruminants*, have these bacteria in their *gastrointestinal* tracts and are, therefore, able to partially use cellulose as a *nutrient* because the bacteria digest it for them.

Cereal grains: See *grains*.

Cerebral: Pertaining to the brain.

Chelation: A *mineral* bound to an *organic* molecule such as a *carbohydrate* or *protein*. If the organic molecule is more readily absorbed than the mineral, chelation will increase the amount of the mineral which can be absorbed. If it is not it will decrease its absorption. For example, *phytate* is a poorly absorbed organic molecule which decreases phosphorus absorption. Chelated minerals are often promoted as being superior to nonchelated minerals. However, their cost with respect to the availability of the mineral chelated must be considered. For example, if chelation doubles the availability of the mineral but triples its cost it would be a poor buy. It would be more economical and the animal would get just as much of the mineral by adding twice as much of the nonchelated mineral.

Chemotherapeutic agents: Chemical substances which inhibit or destroy bacteria or other *infectious organisms,* and which are used in the treatment of infectious diseases. *Antibiotics* are chemotherapeutic agents that are produced by microorganisms.

Chronic: Occurring over a long period of time. Opposite of *acute.*

Cold blooded horses: Horses whose major blood lines are from the original horses of Europe. In Europe two types of horses were developed: the Celtic pony, which was well-formed, light weight, sturdy, and rather short-legged; and the Great Horse of the Middle Ages—the large, slow, and powerful heavy horse. This designation obviously has nothing to do with the temperature of the horse's blood (see *hot blooded horses*).

Colic: Acute abdominal pain. When due to pain from the stomach or intestines, it is referred to as true colic, whereas when due to pain from other organs, such as the urinary tract, liver, reproductive tract, and muscles, it is referred to as false colic. See *strongyles.*

Colon: A portion of the large intestine or terminal portion of the digestive tract which extends from the *cecum* to the *rectum.* In the average size horse, it is 10 to 12 ft (3 to 3.7 m) long and its average diameter is 8 to 10 inches (20 to 25 cm) (Glossary Fig. 2).

Colostrum: The milk secreted by the mare before giving birth and for about the first day following birth. It is high in *antibodies* which protect the newborn against *infectious* diseases. The amount of antibodies in colostrum decreases rapidly after giving birth, as does the foal's ability to absorb them.

Coma: An abnormal state of depressed responsiveness, with absence of response to any stimuli. A state of stupor, or unconsciousness.

Complete feed: A *ration* that contains all of the *nutrients* that are needed by the animal, with the exception of water and salt. The term is generally used for commercially prepared rations that contain both *roughage* and *concentrates.*

Concentrates: A broad classification of feedstuffs that are high in *energy* and low in crude *fiber* (under 18%). Generally considered to be everything in the *ration* not classified as a *roughage,* or the part of the ration composed primarily of *grain.*

Contagious: Communicable or transmissible from one individual to another.

Convulsions: Spasm or violent, involuntary, uncontrollable contraction, or series of contractions, of the muscles.

Cornea: The transparent portion of the front of the eyeball.

Coronary band (coronet): Located at the hair line along the top of the hoof wall, it is the primary source of *nutrients* for the hoof wall which grows from it. Injuries to the coronary band usually leave a permanent defect in the growth of the hoof wall (see Glossary Fig. 1).

Corticosteroids: Substances that affect the animal in the same manner as hormones produced by the adrenal cortex. Usually used to decrease fever and inflammation, and to give the animal a sense of feeling better or not as ill.

Creatinine: A breakdown product of muscle creatine that is excreted in the urine. The amount produced and excreted is directly proportional to the amount of muscle, so the amount excreted daily is fairly constant. The constant quantity of creatinine excreted in the urine can be used to determine the amount of other substances excreted in the urine. The concentrations of all solutes in the urine vary according to the amount of urine excreted, which in turn varies with the amount of water ingested and with kidney function. Therefore, comparing the urine concentration of a substance to the urine concentration of creatinine (the substance:creatinine ratio) eliminates the variable of the urine volume and the need for measuring the amount of a substance excreted over a period of time.

Creep feed, or ration: A *ration* fed to the nursing animal.

Cribbing: A vice in which the horse bites or places its upper incisor teeth on some solid object, pulls down, arches its neck and swallows gulps of air which go into the stomach, not the lungs. Not to be confused with *wood chewing* or *wind sucking* (see Chapter 4).

Crimped: A term for *grain* that is pressed between corrugated rollers to crack the kernels and increase its digestibility (see Cereal Grains, Chapter 2).

Cubes: In equine nutrition the term refers to hay pressed into cubes. These are generally 1 to 2 inches (3 to 5 cm) in size (see Fig. 2–10).

D

Dehy: Meaning *dehydrated.* In animal nutrition it refers to dehydrated alfalfa pellets (see Fig. 2–18).

Dehydrate: To remove all or enough moisture from a feedstuff to prevent its spoiling during storage.

Deworm: Also referred to as worming. Treating an animal to remove stomach and intestinal parasites; in horses these are *ascarids, bots, strongyles, pinworms* and occasionally tapeworms.

Desmotomy: The cutting of ligaments, desmo- meaning ligament, and -otomy meaning to cut.

Diaphragm: A thin muscular membrane or wall separating the thoracic cavity, which contains the lungs, heart, and related structures, from the abdominal cavity, which contains such organs as the stomach, intestines, liver, *spleen*, kidneys, bladder, and, in the female, the *uterus*.

Diet: All *nutrients* consumed by the animal including all feeds, salt, and water, whereas *ration* indicates everything fed to the animal, excluding water and salt.

Digestible: That part of a feed that the animal is able to digest and absorb. Generally considered to be the portion of the feed that the animal is able to utilize.

Distemper: See *strangles*.

Diuretic: A drug that increases the excretion of urine.

Dry matter: The part of the feed which is not water. To determine the amount of a *nutrient* present in the dry matter of a feed, the amount of that nutrient present in the feed is divided by the fraction of the feed which is dry matter. For example, a feed contains 10% moisture and 5% *protein*. Its dry matter content is 90% (100% − 10%) and it contains 5.55% (5% ÷ 0.90) protein in its dry matter.

Dry matter basis: A method of expressing the amount of a *nutrient* in the feed *dry matter* (DM). In the example given above the feed contains 5.55% *protein* on a dry matter basis. If the nutrient content of a feed, its cost, and the like are expressed on a dry matter basis, this value may be converted to the amount present in the feed *as fed* by multiplying the value times the dry matter content of the feed. The feed given above contains 5.55% protein on a dry matter basis and 90% dry matter; therefore, it contains 5% (5.55% × 0.90) protein on an as fed or *as is basis*. One should remember to always multiply or divide by the dry matter value, never the moisture content value, of the feed when converting from *air dry* to a dry matter or to an as fed basis.

$$\% \text{ DM} = 100\% - \% \text{ moisture content}$$
$$\text{amount in DM} = (\text{amount in feed as fed}) \div (\% \text{ DM} \div 100)$$
$$\text{amount in feed as fed} = (\text{amount in DM}) \times (\% \text{ DM} \div 100)$$

Whenever any feed comparisons are made (such as nutrient contents or cost of different feeds, or the horse's requirements as compared to his *ration*) this comparison must be performed with all values expressed on an equal moisture basis. To ensure that this is done, frequently all values are converted to the amount in the feed or ration dry matter.

E

Easy keeper: An animal that requires less feed than others under a similar situation. Opposite of *hard keeper.*

Electrolysis: The breakdown or decomposition of a substance by an electric current. The electric current may be generated by two unlike metals in water, such as zinc and copper.

Electrolytes: Substances that in a solution dissociate into electrically charged particles called ions. Common examples are table salt, or sodium chloride, that in solution dissociates into sodium ions and chloride ions; and hydrochloric acid that dissociates into hydrogen ions and chloride ions. In contrast, a substance such as sand is not an electrolyte because it doesn't dissociate in water into silicone and oxygen. Body salts are electrolytes.

Emaciation: A thin wasted condition of the body.

Emphysema: See *heaves.*

-emia: Attached to the end of a word indicating "in the blood."

Encephalomyelitis: See *sleeping sickness.*

Enema: A liquid introduced into the intestine by way of the anus.

Energy, nutritional: Produced by the animal in the utilization of *carbohydrates, proteins,* or *fats.* These three *nutrients* may be provided in the *ration* or mobilized from the body. When they are in a ration or feed, the total amount of energy they provide is referred to as the energy "content" of that ration or feed, and may be expressed as *calories, therms,* or *TDN.* The concentration of energy in a ration or feed is referred to as the energy "density" of that ration or feed.

Ensilage: See *silage.*

Epiphysis: The place where bone growth occurs.

Epiphysitis: Inflammation of the *epiphysis.* Incorrectly, but commonly, used to describe conditions occurring in the growing horse due to inflammation of the *metaphysis.*

Epistaxis: Bleeding from the nose. In the horse this may be caused by trauma, such as from a tube passed through a nostril into the stomach, by blood coagulation disorders, or by blood coughed up from broken blood vessels in the lungs resulting most often from *chronic bronchitis* and *heaves* (see Chapter 4).

Ergot: 1. A fungus (Claviceps spp) which grows under warm, moist conditions affecting and finally replacing the seeds of *cereal grain* plants, especially rye, oats, and wheat, and the grasses Tobossa, Bahia, Dallis, and Kentucky bluegrass. Ergot is small, hard, and dark brown to black in color, resembling dried mouse feces. It causes constriction of the *arteries,* decreasing blood flow to the tissues. If blood flow is decreased sufficiently the tissue dies and becomes *gangrenous*. The major tissues affected are the extremities—ears, tail, and limbs. Instead of the gangrenous form other effects may be abortions or nervous system disorders, including excitability, trembling, incoordination, *tetany*, and death.

2. A small mass of horn in the tuft of the hair at the back of the horse's *fetlock* is called the ergot (see Glossary Fig. 1).

Ether extract: Fatty substances of feeds that are soluble in ether or other *fat* soluble solvents. A measure of the crude fat content of the feed.

Euthanasia: Putting an animal to death painlessly.

Exertion myopathy: See *azoturia*.

Expectorant: A medicine which aids in the expulsion of mucus or exudate from the lungs, *bronchi*, and trachea.

F

Fats: Chemically, fats are glycerol (the 3 carbon compound shown below) plus fatty acids, all of which are composed of carbon, hydrogen and oxygen. Fats are necessary in the *diet* for the absorption of substances which are soluble only in fats, such as *vitamins* A, D, E, and K, and as a source of fatty acids needed for body structure. Most dietary fatty acids contain 16 or 18 carbon atoms.

$$
\begin{array}{c}
\text{H} \\
| \\
\text{H—C—O—fatty acid} \\
| \\
\text{H—C—O—fatty acid} \\
| \\
\text{H—C—O—fatty acid or a carbohydrate} \\
| \\
\text{H}
\end{array}
$$

Fat or diglyceride (2 fatty acids) or
Triglyceride (3 fatty acids)

Ferritin: A *protein* that binds iron for storage in the body, primarily in the liver and *spleen*. A small amount is present in the blood. Its concentration in the blood is directly proportional to the amount of iron in the body.

Fetlock: The horse's ankle joint (see Glossary Fig. 1).

Fiber or crude fiber: The portion of a feed high in *cellulose* and lignin, which are poorly digestible.

Fines: Substances that are small enough and heavy enough to settle to the bottom of a feed.

Flexor tendons: The tissue connecting a muscle to a bone which flexes the extremity or leg.

Floating the teeth: Filing off sharp points of enamel that develop on the teeth. These sharp points may lacerate the tongue and cheeks (see discussion on care of teeth, Chapter 10).

Flu: See *influenza.*

Fodder: Coarse feeds, such as corn or sorghum stalks, with the *grain* removed.

Forage: Plant material consumed by animals that is high in *fiber* (greater than 18%); the stems and leaves of plants, with or without the seeds or *grain.* Includes pasture, hay, *straw,* and *silage.*

Founder (laminitis): An inflammation of the laminae of the horse's foot. Laminae are located between the bone and the hoof, and contain blood vessels which nourish the hoof. When inflamed, the laminae swell between these two rigid structures causing pressure, pain, and tissue damage. This results in separation of the hoof wall from the laminae, and downward rotation of the third phalanx (most distal bone of the foot), so that in extreme cases it may penetrate the sole. Founder may be caused by (1) ingestion of excessive amounts of *grain,* or large amounts of cold water by a hot horse or, in some horses (particularly pony breeds), of lush green *forage;* (2) *infectious* conditions such as severe pneumonia, or *uterine* infection following foaling; or (3) concussion to the feet from running on a hard surface (see Chapter 4). There have also been reports of horses foundering after "stress" conditions (accidents).

Free-choice: Refers to feed or water being available for the animal to eat or drink as much and as often as it wants.

Fungal spores: Small reproductive elements given off by fungi which, when inhaled by the horse, can cause *heaves, bronchitis,* and coughing.

G

Gangrene: Death of large amounts of body tissue.

Gastrophilus: See *bots.*

Gastrointestinal: Gastro- meaning stomach, and intestinal meaning intestines or guts.

Glucose: A simple sugar or monosaccharide that is *metabolized* by the animal for *energy.* It is the major source of energy for the brain.

Glycogen: A storage form of *glucose* present primarily in the muscle and liver.

Goiter: Enlargement of the thyroid gland, causing a swelling at the throatlatch. It may be caused either by a deficiency or an excess of iodine in the *diet,* or by the ingestion of large enough quantities over a period of time of substances that interfere with the thyroid function, e.g., kale, white clovers, cabbage, rutabaga, and turnips.

Thiouracils, thiourea, and methimazole are drugs which also may cause goiter (see Fig. 1–11).

Grain: The seed of plants used for food, e.g., corn (maize), milo, oats, wheat, barley, and rye.

Graze: To consume standing vegetation.

Grease heel: A bacterial or fungal infection of the skin on the heel and back of the *pastern* that may occur as a result of the horse's standing in moist, dirty bedding (see Fig. 10–7).

Groats: Cereal grain kernels after removal of the *hulls,* i.e., dehulled cereal grain.

H

Hard keeper: An animal that is nutritionally unthrifty and requires more feed than others under a similar situation. Opposite of *easy keeper.*

Hay belly (grass belly): Said of a horse with a distended abdomen due to the presence of excessive amounts of bulky feeds in the intestinal tract.

Haylage: Feed from grass or leguminous plants that are cut, not allowed to dry and then ensiled (see *silage*).

Heating feed: A feed in which much of its *energy* is given off as heat during its digestion, absorption, and utilization. In contrast to popular belief, corn is not a heating feed (see Cereal Grains, Chapter 2).

Heaves (*pulmonary emphysema* or broken wind): A respiratory disturbance resulting from reduced elasticity and rupture of the small sacs (alveoli) in the lungs where gas exchange between the blood and inspired air takes place. Requires forced expiration of air from the lungs.

Heave line: A ridge of abdominal muscle that has increased in size because of the forced expiration required when *heaves* is present. In order to push air out of the inelastic lungs, it is necessary to contract the abdomen and push the intestines against the *diaphragm.* The ridge runs in a straight line from the middle of the flank diagonally forward and down to the rib cage toward the point of the elbow (see Fig. 4–6).

Hematocrit: The percent of the blood that is *red blood cells.* Also called the packed cell volume, since the hematocrit is measured by centrifuging the blood and measuring the red blood cells that the centrifugal force has packed into the bottom of the tube.

Hemoglobin: See *red blood cells.*

Hemolysis: The breakdown or destruction of *red blood cells.*

Hemolytic icterus in foals (neonatal isoerythrolysis or "dishrag foals"): Affected foals are normal and healthy at birth. They nurse, and are active for a short period. After 12 to 36 hours they become dull, sluggish, and weak, stop nursing, and may go down. Heart rate and respiratory rate are increased, particularly after exertion. Membranes are pale for the first 24 hours, then become yellowish. Later the urine may become blood tinged. The course of the illness is 1 to 10 days; most foals die at 3 to 4 days of age. The cause is the ingestion of *colostrum* that contains *antibodies* that destroy the foal's *red blood cells* (see Chapter 7).

Herbivores: Animals that live on or prefer plant materials, in contrast to *carnivores* who live on or prefer animal tissues. *Omnivores* are animals that can live on either animal or plant materials. In reality all animals are nutritionally omnivores, and can utilize both animal and plant materials. However, some require or prefer more plant material, and others, more animal material. Progressing from animals that are more herbivorus to those that are more carnivorus would be cattle, horses, pigs, humans, dogs, and cats.

Hives: Small swellings or welts on the skin. They appear suddenly, and are caused by an *allergic* response to such things as feed, insect bites, drugs, and insecticides. When caused by feed, they may be referred to as *protein bumps* (see Fig. 1–4).

Hock: The tarsal joint in the hind legs located about one-half way between the ground and the abdomen (see Glossary Fig. 1).

Homeostatic: A tendency to uniformity or stability in the normal body states.

Hot blooded horses: Horses that originated from Arabians, Barb, and Turkmene horses. These horses came from northern Africa, along the Mediterranean coast, and the Arabian peninsula. Breeds which have originated at least in part from them include Thoroughbreds, Standardbreds, the American Saddle Horse, the Morgan, the Quarter Horse, and the Tennessee Walker. This designation obviously has nothing to do with the temperature of the horse's blood (see *cold blooded horses*).

Hot feed: To some this means a feed high in *energy*. If the feed provides more energy than the horse can use and the horse feels good, it has a tendency to make the horse spirited. Another meaning for hot feed is that which produces a great deal of heat. This will keep the animal warmer in cold weather, but causes increased sweating and discomfort in warm weather.

Hulls: Outer covering of *grain*.

Hyper-: A prefix meaning above normal, e.g., hyperparathyroidism indicates excessive production of hormone by the *parathyroid* gland.

Hyperplasia: The enlargement or overgrowth of a part of the body due to an increase in the number (not size) of its cells.

Hypertrophy: The enlargement or overgrowth of a part of the body due to an increase in size (not numbers) of its cells.

Hypo-: A prefix meaning below normal.

I

Icterus (jaundice): Excess bile pigments in the blood, or deposited in the skin and membranes giving them a yellow color. May be caused by liver damage or excessive destruction of *red blood cells (hemolysis).* See *hemolytic icterus in foals.*

Immunoglobulins (antibodies): Proteins found in body fluids that produce an immunity or resistance to foreign substances such as bacteria or viruses by combining with them and helping to inactivate them.

Infectious organisms: Any organisms, such as bacteria, viruses, protozoa, or fungi, that are capable of invading, or entering, the body and causing disease. The resulting infectious disease may or may not be *contagious.*

Influenza (flu): A highly *contagious* viral disease caused by at least two distinct strains of virus. It affects horses of all ages but is most common in horses less than three years old. Following inhalation of the virus, susceptible horses show clinical signs within 1 to 3 days. It is characterized by fever (up to 106°F or 41°C), depression, often violent coughing, decreased appetite, and usually, a mucoid nasal discharge early in the disease (Glossary Fig. 8). In the absence of secondary complications, the respiratory lining regenerates and recovery occurs in approximately three weeks. The most common secondary complications of viral respiratory diseases, such as influenza or rhinopneumonitis, are the development of a bacterial pneumonia resulting in a persistent, often productive cough, fever, leukocytosis, and a purulent nasal discharge. Prolonged, excessive coughing may predispose the horse to pulmonary emphysema or hemorrhage, and bronchitis. Additional secondary complications that may occur include pharyngitis, diarrhea, pleuritis, pleural effusion, and heart disease. Few horses are able to return to athletic usefulness after developing pleuritis. Occasionally, in young, old, debilitated, or stressed horses, these viruses may localize in the muscle, causing myositis, or in the heart, causing myocarditis. Myocarditis may result in arrhythmias such as atrial fibrillation. Treatment of viral respiratory disease, no matter what disease is responsible, influenza or rhinopneumonitis, is mainly rest and supportive therapy. Complete rest for 3 weeks is the most important aspect in the treatment of viral respiratory diseases and is strongly recommended in order to prevent secondary complications. With rest, other therapy may not be needed. Without adequate rest, other treatments are generally ineffective. Antibiotics are ineffective against viruses, however, adequate doses of procaine penicillin G (a minimum of 10,000 IU/lb or 4500 IU/kg of body weight injected intramuscularly twice a day for 3 and preferably for 5 days) may be beneficial in protecting the horse against secondary bacterial infection. Inadequate doses for inadequate time periods predispose the horse to a resistant bacterial infection. Many veterinarians advocate the use of anti-inflammatory drugs such as corticosteroids or Butazolidin. Corticosteroids may in some cases be used once or twice, but repeated administrations are contraindicated since they suppress the animals' immunity and, therefore,

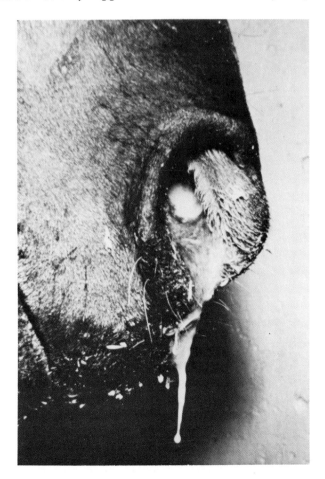

Glossary Fig. 8.   Mucopurulent nasal discharge typical of that from the horse with influenza.

increase the risk of the development of bacterial pneumonia and pleuritis. Horses with respiratory disease should be immediately isolated from other horses, whenever possible. Every effort should be made to minimize contact with other horses. Separate water, feed buckets, halters, and grooming equipment should be used for each horse. Barn personnel should handle sick animals last, since they may transmit the viruses to unaffected horses. During an outbreak, all horses not ill should be vaccinated to limit the spread of the disease. There is no evidence that vaccination of a horse incubating a viral respiratory disease is harmful, however, clinically ill horses should not be vaccinated. See the section on vaccination in Chapter 10 for prevention of respiratory diseases, such as influenza and rhinopneumonitis.

Inoculate: see *vaccinate.*

Inorganic: Noncarbon-containing compounds such as *minerals.*

Intra-: A prefix meaning into or in, e.g., *intra*muscular, meaning into or in the muscle, or *intra*venous, meaning into the *vein.*

Isoantibodies: *Antibodies* or *immunoglobulins* produced in the body against the body's own cells or tissues; e.g., anti- *red blood cell* isoantibodies are antibodies against the animal's own red blood cells, which destroy these cells.

-itis: Attached to the end of a word indicating inflamed, e.g., *laminitis* indicating inflammation of the laminae of the foot.

J

Jaundice: See *icterus.*

Jugging: Jugging a horse is a practice in which 500 ml (1 pt), or more, of a sterile solution of *amino acids, electrolytes, vitamin* B complex, and frequently *glucose* are given into the jugular *vein.*

K

Kilocalorie (kcal): See *calorie.*

L

Labile: Unstable, easily destroyed.

Laminitis: See *founder.*

Legumes: Plants, such as alfalfa (lucerne), clovers, birdsfoot trefoil, lespedeza, vetches, and peas that obtain nitrogen through bacteria that live in their roots. The plant converts the nitrogen to *protein,* so that legumes are higher in protein than nonlegumes such as grasses.

Ligament: A tough, fibrous band which connects bones or supports viscera.

Lipids: See *fats.*

Lockjaw: See *tetanus.*

Lucerne: Alfalfa, a *legume roughage.*

Lysine: The *amino acid* which is most deficient in the growing horse's *ration,* the deficiency of which slows growth rate (see Chapter 1).

M

Macro-: Meaning large, e.g., *macro*minerals meaning *minerals* needed in large quantities. See minerals. Opposite of *micro-.*

Maintenance ration: A *ration* that is adequate to prevent any loss or gain of weight when the animal is at rest in an environmental temperature in which additional *energy* is not required to heat or cool the body.

Maize: Corn, a cereal *grain.*

Meconium: Feces accumulated in the bowels prior to birth.

Megacalorie (Mcal): One Mcal equals 1000 nutritional *calories* or *kilocalories* and 1 *therm.*

Metabolism: Chemical reactions occurring in the body. Refers to all changes in or utilization of a *nutrient* that takes place following its absorption from the intestine.

Metacarpal: The front leg *fetlock* joint, or the cannon bone and splint bones on the front legs (see Glossary Fig. 1).

Metaphysis: A zone of spongy bone between the *epiphysis* and the rest of the long bone. Where growth of the long bone occurs.

Metatarsal: The hind leg *fetlock* joint, or the cannon bone and splint bones of the hind legs (see Glossary Fig. 1).

Methemoglobin: Hemoglobin is the iron-containing compound in *red blood cells* that binds with oxygen so that it can be transported to the cells. When its iron is converted from the ferrous to the ferric state, it is called methemoglobin and can no longer combine with oxygen.

Micro-: Meaning small, e.g., *micro*scopic or *micro*minerals. See *minerals.* Opposite of *macro-.*

Minerals (ash): Noncarbon-containing *(inorganic)* elements such as salts, calcium, phosphorus, and magnesium. The total amount present is determined by burning the carbon-containing *(organic)* matter and weighing the residue, which is called ash. This is a relatively meaningless value unless the amounts of the specific substances making up the mineral or ash are given. In nutrition, minerals are classified into two groups according to the amount needed by the animal; *macro-* meaning large amounts, and *micro-* meaning small amounts. Macrominerals are sodium chloride (common salt), magnesium, potassium, calcium, phosphorus, and sulfur, and are all needed in the *ration* in a number of parts per hundred, or percent. Microminerals are iodine, molybdenum, manganese, iron, cobalt, zinc, copper, and selenium, and are all needed in the ration in a number of parts per million (ppm or mg/kg). Microminerals are also referred to as *trace minerals*. With the exception of iodine, which is a necessary constituent of thyroid hormones, the other trace minerals are necessary primarily as components of enzymes. Enzymes are required for many of the chemical reactions that occur in the body.

mM (Millimole): A unit of concentration that indicates the number of molecular weights of a substance in milligrams in one liter of water. For example, the molecular weight of sodium is 23, therefore, 23 mg of sodium in one liter would be 1 mM of sodium.

mOsm (milliosmoles): 1/1000 osmoles. See *osmolarity*.

Mycotoxins: Substances produced, particularly under warm, moist conditions, by fungi or molds, which cause deleterious effects to biological systems. Most *antibiotics* are mycotoxins that are more toxic for bacteria than for animals, and are therefore given to animals to kill *infectious* bacteria. Many mycotoxins are harmful to the horse and cause a wide variety of symptoms. General symptoms often attributed to mycotoxins are decreased feed consumption, infertility, poor hair coats, decreased performance and growth rate, abortions, diarrhea, liver and kidney damage, nervous system affects (such as tremors, incoordination, and hyperexcitability), hemorrhage, and death. Usually symptoms don't occur until levels of 100 to 300 ppb are present in the *ration*, although this depends on the mycotoxins involved. Except for fatalities, the effects of most of the mycotoxins are reversible once the contaminated feed is removed from the ration. There are 25 to 30 different mycotoxins that may be detrimental to the horse. The best known causes of mycotoxicosis are (1) aflatoxins, ochratoxins, and tremortin A produced by Aspergillus and Penicillium, (2) trichothecene (T-2) produced by Fusarium and Myrothecium, (3) fescue toxicity, which may be caused by a mycotoxin produced by Fusarium tricinctum (see Chapter 4), (4) *ergot* produced by Claviceps spp., (5) zearalenone (F-2) produced by Fusarium, (6) salframine produced by Rhizoctonia, (7) citrinin and rubratoxin produced by Penicillium, and (8) sweet clover toxicity, caused by the mycotoxin dicoumarin which prevents blood coagulation, resulting in hemorrhage, and is produced by Aspergillus and Penicillium sp. in improperly cured sweet clover hay. A definitive diagnosis of mycotoxicosis is often difficult. Some veterinary diagnostic laboratories are able to analyze for a number of the major mycotoxins. However, portions of a feed may contain high levels of mycotoxins, and in other portions, none may be present. Thus, numerous samples of the suspected feed are necessary. Factors useful in making a presumptive diagnosis include (1) recognition of clinical signs that occur with mycotoxicosis, (2) failure to isolate infectious agents or respond to antibiotic therapy, (3) presence of moldy feed, or environmental factors favorable for fungi development and mycotoxin production (usually warm, moist conditions), (4) indications are decreased feed *palatability*, or feed refusal, and (5) test feeding of animals with the suspected feed, remembering that the young are generally the most susceptible and that several weeks may be required before symptoms occur. The best way to avoid mycotoxicosis is to never feed moldy feeds, although not all moldy feeds contain mycotoxins and mycotoxins may be present in feeds that are not visibly moldy.

N

Nebulization: Generally pertains to a treatment in which a substance is converted to a spray so that it can be inhaled; e.g., a vaporizer nebulizes water, and drugs which may be included in the water.

Glossary Fig. 9. Nits, or eggs, attached to the horse's hair by botflies. Compare these to those from lice (Glossary Fig. 10). When the horse licks or scratches itself with its teeth, botfly nits enter the mouth and are swallowed. In the stomach they develop into bot larvae (Glossary Figs. 6 and 7), which remain there until spring, when they are passed in the feces. The larvae then pupate to become botflies (Glossary Fig. 5).

Glossary Fig. 10. Nits or lice eggs. Compare to those laid by botflies (Glossary Fig. 9).

Necropsy (autopsy): A postmortem, or after-death, examination.

Necrosis: Death of body tissue or cells.

Negative nutrient balance: Refers to the state when more of a nutrient is being lost from the body than is being taken in. Opposite of a positive nutrient balance.

Nits: Eggs laid on the horse's hair by *bot*flies (Glossary Fig. 9) or lice (Glossary Fig. 10). They are yellow and are about the size of the head of a pin.

NFE (Nitrogen-Free Extract): The portion of a feed consisting principally of soluble *carbohydrates*, e.g., sugars and starches. The most digestible and utilized carbohydrate fraction of the feed. Used primarily as a source of *energy*. The percentage of NFE in a feed is determined by subtracting the sum of the percentage of moisture, crude *protein*, crude *fiber*, and *ash* from 100.

Nutrients: Ingested substances necessary for the support of life. The chief classes of nutrients are *carbohydrates, fats, proteins, minerals, vitamins,* and water.

O

Oil: Fats that have a melting point below normal environmental temperatures.

Omnivores: See *herbivores.*

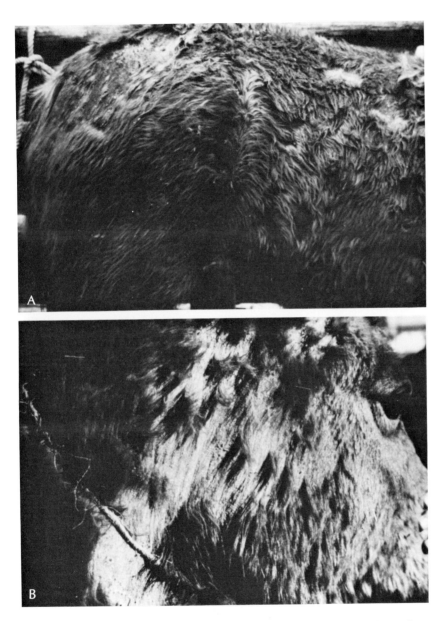

Glossary Fig. 11(A,B). Hair loss due to lice. Lice may also cause extensive blood loss and poor condition.

Open knees: A concave appearance to the front of the knee or *carpus* caused by *epiphysitis* occurring just above the knee.

Oral: Pertaining to the mouth.

Organically-complexed: See *chelation*.

Organic: A substance containing carbon.

Osmolality: An indication of the number of particles dissolved in a fluid. Water moves across membranes in the body to try to equalize the osmolality, or number of particles in solution, on each side of the membrane. Thus, if the osmolality outside the cells is higher than that inside the cells, water will move out of the cells. This increases the osmolality of the fluid inside the cells, and decreases it outside the cells, i.e., equalizes the osmolality on each side of the cell membrane.

Osteoarthritis: Inflammation of the bone and joint. Osteo- pertains to bone, arth- pertains to a joint, and -itis indicates inflammation.

Osteochondritis: Inflammation of both bone and cartilage. Osteitis is inflammation of the bone and chondritis is inflammation of the cartilage.

Osteochondritis dissecans: The presence of a loose piece of cartilage or fragment of cartilage and bone present in the joint. It usually occurs in horses less than 18 months of age. Joint capsule distention and moderate to severe lameness are common effects. It most commonly occurs in rapidly growing horses, often described as the best in the group, being "pushed" nutritionally by feeding high amounts of *concentrates*. Surgery to remove the loose flap of material is usually necessary. The method of prevention is to slow growth rate by feeding less *grain*, while feeding a *ration* that meets all of the animal's nutritional requirements.

Oxalate: A substance that binds positively charged *minerals (cations)*, such as calcium, decreasing their absorption from the *ration*. After absorption, oxalates bind calcium in the bloodstream forming a precipitate which is deposited in the kidney, and may thereby cause kidney failure.

Oxyuris equi: See *pinworms*.

P

Palatability: The desirability of a food for ingestion.

Parascaris equorum: See *ascarids*.

Parathyroid gland: Small glands located on or near the thyroid gland which is in the vicinity of the throatlatch. A decrease in blood calcium concentration stimulates these glands to secrete a substance (parathyroid hormone) that increases intestinal calcium absorption, calcium, and phosphorus mobilization from the bone, and phosphorus excretion in the urine. It also decreases urinary calcium excretion. Its function is to prevent a decrease in the blood calcium concentration.

Parturition: The act of giving birth.

Pastern: The area between the *fetlock* and hoof (see Glossary Fig. 1).

Phytate: A six-carbon ring with a phosphate molecule bonded to each carbon atom. Because of this attachment, its phosphate is poorly available to the animal. The phosphate of phytate binds positively charged *minerals (cations)*, such as calcium, decreasing their absorption.

Pinworms (Oxyuris equi): Adult pinworms are found mainly in the *colon* and *rectum* of the horse (see Glossary Fig. 2). They may cause itching and restlessness. Affected horses may rub the tail on any stationary object, and may wear away the hair at the base of the tail and rump. See discussion of intestinal parasites in Chapter 10 for treatment.

Placenta: The organ in contact with the mother's *uterus;* it is attached to the fetus by means of the *umbilical* cord. It is passed after birth and is, therefore, referred to as the "after-birth."

Plasma: See *red blood cells.*

Pulmonary emphysema: See *heaves.*

Protein: A nitrogen-containing *organic* (carbon-containing) compound, made up of many *amino acids* that are hooked together (see amino acids). It is necessary in the *diet*, for the body to use in producing all of its tissues. It may also be used by the animal for *energy.* Crude protein indicates the total amount of protein in a feed; when only protein is stated, crude protein is generally what is meant. Digestible protein indicates only that protein that can be digested, and can therefore be used by the animal. Most protein in natural feeds for the horse is about 75% digestible. Thus if a feed contains 10% crude protein it would contain about 7.5% digestible protein. If the animal needs 16% crude protein in the *ration*, it would need 12% digestible protein (16% × 0.75).

Protein bumps: See *hives* and Figure 1–4.

Protein supplements: Feeds that may be added to a *ration* to increase its protein content.

Protein quality: A term used to describe the *amino acid* balance of a protein. A protein is good, or high-quality, when it contains all of the amino acids in proper proportions needed by an animal, and is poor, or low-quality, when it is deficient in either content or balance of the amino acids needed.

R

Radiographs: Photographs made by using radioactive rays. The radioactive rays are often referred to as x-rays.

Radius: The major bone in the forearm (see Glossary Fig. 1). There are two bones in the forearm, the radius and the ulna. They are united in the horse.

Ration: The feed supplied or available to an animal.

Rectum: The terminal end of the large intestine, or digestive tract, which extends from the *colon* to the anus (see Glossary Fig. 2).

Red blood cells: The blood consists of a noncellular liquid portion called *plasma* and the cells suspended in it. The greatest number of cells are red blood cells. These cells contain the iron-containing compound, *hemoglobin.* Hemoglobin binds oxygen in the lungs and releases it to the cells throughout the body.

Renal: Refers to the kidneys.

Renal Clearance Ratio: The amount of a substance excreted in the urine with respect to the amount of *creatinine* excreted. The amount of creatinine excreted is fairly constant. Thus, comparing the amount of a substance excreted to the amount of creatinine excreted eliminates dilutional effects on that substance's urine concentration and the necessity of doing a quantitative collection of urine over a period of time (see Chapter 9).

Rhinopneumonitis: A disease caused by two major types of equine herpesviruses which cause three disease syndromes in horses: acute upper respiratory infection, abortion, and central nervous system infection. It is reported to be the cause of 30 to 40% of all respiratory disease in horses. The respiratory disease caused by rhinopneumonitis is characterized by fever (up to 106°F or 41°C) for 1 to 7 days, mild cough, and a serous, watery nasal discharge. Appetite may be decreased or may remain unaffected. Infrequently, diarrhea and stocking-up may occur. With proper treatment, recovery is usually complete within 1 to 2 weeks in uncomplicated cases (see glossary description of influenza). In the pregnant mare, rhinopneumonitis may cause abortion. Abortion most commonly occurs during the eighth to eleventh month of gestation, and is characterized by complete and rapid expulsion of the infected fetus and membranes. Less commonly, instead of abortion, infected mares may give birth to a live, but weak, foal, which usually dies of acute pneumonia or pleuritis in the first few days of life. Infections that result in abortion or birth of weak foals usually occur at the same time as an outbreak of respiratory disease in young horses on the farm. Affected mares may never show any signs of respiratory disease. Rarely, there may be infection of the central nervous system, resulting in incoordination frequently associated with urinary incontinence, a decrease in tail tone, and variable anesthesia of the perineal area. In milder cases of central nervous system infection, the horse may recover. Some horses have persistent neurologic deficits; however, when infection of the central nervous system occurs it is usually fatal. See the section on vaccination in Chapter 10 for the prevention of rhinopneumonitis.

Ringworm (Dermatomycosis): A fungal infection of the skin characterized by sharply demarked, round, gray, dry, scaly areas of hair loss (see Fig. 10–8). It occurs most commonly in younger animals.

Roughage: Feed consisting of bulky and coarse plants or plant parts, with a high *fiber* content and low total digestible *nutrients*, arbitrarily defined as feed with over 18 percent crude fiber.

Rumen: Often referred to as the first stomach of cud-chewing animals such as cattle, sheep, goats, deer, and antelope. However, it is not a true stomach but a large compartment which, when full, comprises 15 to 20% of the animal's body weight. It is the first compartment ingested foods reach after being swallowed; next is the reticulum, then the omasum, followed by the abomasum, small intestine, *cecum*, large intestine, *rectum*, and anus. The abomasum is the true stomach and is similar to that of other animals. The rumen contains many bacteria that digest much of the feed that the animal ingests. It serves many of the same functions as the cecum in the horse.

Ruminant: An animal that has a *rumen*.

S

Sclerosis: A hardening, especially from inflammation.

Serum: The noncellular liquid remaining after blood clots.

Seedy toe: A separation of the hoof wall from the sole of the foot leaving a crack. Organisms may enter the foot through this crack and cause foot infections and *abscesses*. It occurs most commonly as a result of *founder*.

Silage (ensilage): Fermented *forage* plants. Corn, sorghum, and small *cereal grains* are cut and not allowed to dry, or moisture is added to them. The entire plant (stalks, leaves, and grain) is chopped and packed into a facility which protects it from exposure to air. Permits maximum yield of *nutrients* from a unit of ground.

Sleeping sickness (*encephalomyelitis, encephalitis, brain fever, or blind staggers*): A disease caused by three major types of virus: Eastern, Western, and Venezuelan strains. The virus is transmitted by blood sucking insects, such as mosquitos, from birds and rodents to the horse or humans. The bird and rodent are unaffected by the virus, as are animals other than the horse and humans. The virus is not transmitted from horses or humans to other horses or humans, i.e., a horse cannot get the disease from another horse. Symptoms include high fever, incoordination, drowsiness, partial loss of vision, grinding of the teeth, reeling gait, inability to swallow, and, in later stages, paralysis. Once symptoms occur it is usually fatal regardless of treatment. The disease is prevented by vaccinating twice, two to three weeks apart, with annual revaccination several weeks before mosquito season begins.

Spleen: A glandular organ located near the stomach. It is a flattened oblong shape that varies in size, but in the horse is about 10 by 20 inches (25 by 50 cm). It is dark purple with a soft consistency. It serves as a reservoir for *red blood cells,* and assists the body in combating *infectious organisms.*

Splints: Boney enlargement occurring at the attachment of the splint bone to the cannon bone (see Glossary Fig. 1). The cannon bone is the large bone between the knee *(carpus),* or *hock,* and the ankle *(fetlock).* The splint bones are small, pencil-sized bones located on each side of the cannon bone. Splints most often occur on the inside of the forelegs several inches below the knee. They are caused by hard training, poor conformation, trauma, or improper nutrition in the young horse (one to four years old). Initially, there is lameness and later, a hard boney enlargement.

Stifle (femoro-tibial joint): A joint in the hind legs located at the level of the bottom of the abdomen (see Glossary Fig. 1).

Stocking up: The accumulation of excess fluid in the tissues (edema) under the skin causing a diffuse swelling of the area (Glossary Fig. 12). Most commonly occurs in the legs, but may also occur in the prepuce and underline (see Glossary Fig. 1). The most common cause is sudden inactivity, such as putting a horse in a stall, particularly following a period of physical activity. Turning the horse out in a paddock, or exercising a few hours a day, will prevent stocking up. Applying bandages to keep pressure on the legs will also prevent fluid accumulation. Stocking up not associated with inactivity may be due to congestive heart failure, or more commonly, a decrease in the concentration of *proteins* in the *plasma.* This may be caused by a decrease in protein synthesis because of liver damage, or excessive losses of proteins from the body such as in blood loss or some forms of diarrhea. High protein *diets* and alfalfa will not cause or predispose the horse to stocking up. A protein deficient diet may. However, the horse would have to be in extremely poor condition before this occurred.

Stover: Mature, sun-cured stalks of corn or milo from which the seeds or *grain* have been removed.

Strangles (distemper): A disease caused by the bacteria Streptococcus equi and characterized by fever, depression, nasal discharge (Glossary Fig. 13), cough, and swollen lymph glands (particularly noticeable in those under the jaw) (Glossary Fig. 14). Most

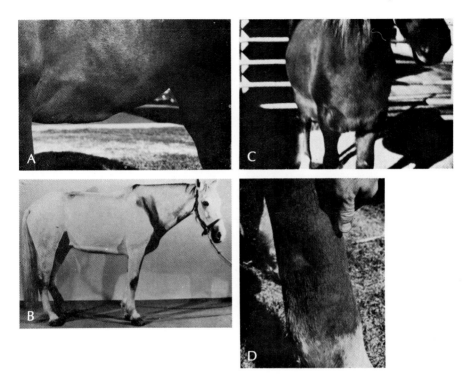

Glossary Fig. 12(A,B,C,D). Stocking up, or subcutaneous edema. Accumulation of excessive fluid under the skin. The fluid accumulates in the lower aspects of the legs, along the underline, and prepuce (Glossary Fig. 1). This causes a diffuse swelling in those areas. Note the edema and swelling along the underline of the two horses in A and B, in the brisket of the horse in C, and around the cannon bone and down the leg in the horse in D.

commonly occurs in younger horses, but may occasionally occur in the older horse. The usual site for strangles is in the throat, but in some cases the internal organs are involved, resulting in a condition called "bastard" strangles. In strangles, *abscesses* frequently develop and rupture, expelling a thick, yellowish pus either to the outside or into the body; rupture into the body may result in death (see Glossary Fig. 3). Prevention involves vaccination twice, 2 to 3 weeks apart beginning preferably at 8 to 12 weeks of age. Annual revaccination should be given for at least the first few years of life. Animals incubating strangles may have a severe reaction to vaccination, so one should vaccinate with caution in the face of an outbreak. Vaccination is frequently ineffective then.

Straw: Small *cereal grain* plant residue, stems, or stalks, after removal of the *grain* or seeds.

Stroke volume: The amount of blood the heart pumps with each beat.

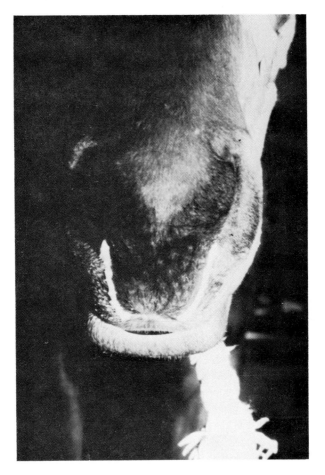

Glossary Fig. 13. Yellowish-white nasal discharge, typical of the horse with strangles (distemper).

Strongyles (blood worms): There are several different types of strongyles. They can affect all horses of all ages. Those that are the most harmful to the horse are the large strongyles. Large adult strongyles are about 1 inch (2.5 cm) long and range from red to gray (Glossary Fig. 15). Adult worms live on blood sucked from the intestine, and they lay eggs that are passed in the manure (Fig. 10–11). After the horse swallows the larvae that develop from these eggs (Glossary Fig. 16), the larvae migrate through the wall of the intestine and into the walls of the arteries. This may result in fibrous thickening of the arterial wall and decreased blood supply to the tissues and organs supplied by that *artery* (Glossary Fig. 17). It may decrease the blood supply to the back legs,

Glossary Fig. 14. Swollen lymph nodes and throatlatch in a horse with strangles (distemper).

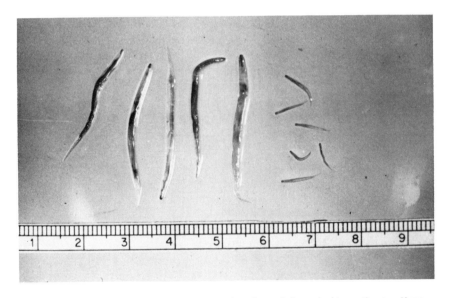

Glossary Fig. 15. Large and small adult strongyles. The scale is marked in centimeters (1 cm = 0.4 inch). Adult strongyles lay eggs which are passed in the manure (Figs. 10–10, 10–11). These eggs hatch into larvae (Glossary Fig. 16).

Glossary Fig. 16. Strongyle larvae in a drop of moisture on a blade of grass. These larvae develop from eggs passed in the feces (Figs. 10–10, 10–11), and are ingested by the horse. After ingestion, they migrate into the walls of the arteries, affecting primarily those that supply blood to the intestines (Glossary Fig. 17) and hind legs.

causing increased fatigue of the hind legs. In addition, it is the predisposing cause of over 90% of all cases of *colic* in the horse. Many factors, such as a rapid change in the type, quantity, or amount of feed ingested, stress, excessive cold water ingestion by a hot horse, lack of water, ingestion of sand or dirt, and physical activity resulting in intestinal displacement, are recognized as being immediate causes of colic. However, in most cases these factors would not cause colic if the pre-existing strongyle-induced damage was not present. Strongyles also suck blood and may cause *anemia*, weakness, and poor condition. Some strongyle larvae migrate through the liver and kidney capsule and may damage these organs (see discussion of intestinal parasites, Chapter 10 for treatment).

Stubble: The basal portion of plants remaining above the ground after the top portion has been harvested.

Supplement: A feed or feed mixture added to a *ration* to increase the amount of a specific *nutrient(s)*.

Sweet feed: Refers to a *concentrate* containing molasses.

Synchronous Diaphragmatic Flutter (SDF, or thumps): Contraction of the *diaphragm* in synchrony with the heart. Clinically, it is

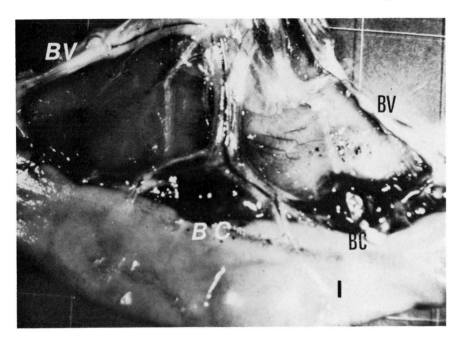

Glossary Fig. 17. Thrombi or blood clots (BC), and the damage they cause to the arteries that supply blood to the intestine. The blood clots are the two black areas in the center and center right side of the figure. Three blood vessels (BV) are shown radiating down from the top center of the figure to the intestine (I), which passes across the width of the bottom of the figure. Damage of this type is the predisposing cause of over 90% of all cases of colic in the horse.

manifested by sudden, bilateral, and occasionally unilateral, movements of the horse's flanks, and sometimes hind legs, every time the heart beats. It is caused by a decrease in the *plasma*, calcium, chloride, and/or potassium concentration due to a loss of these *electrolytes* from the body in the sweat and urine as a result of physical activity. Prevention and treatment is to give these electrolytes during and/or after physical activity (see Table 5–3).

T

Tapeworms (Anoplocephala): A type of worm present in the intestine of 10 to 67% of all horses in the United States and Canada. They utilize nutrients ingested by the horse, but usually are not a widespread problem. However, when they cause a problem, it occurs acutely. Clusters of the worms obstruct the ileocecal sphincter (where the small intestine enters the cecum—see Glossary Fig. 2) resulting in colic and death. The drug pyrantel pamoate at a dose of 3 mg/lb (6.6 mg/kg) body weight is effective in treating and preventing tapeworms.

TDN: See *total digestible nutrients.*

Tetanus (lockjaw): A disease caused by the bacterium Clostridium tetani which is widely distributed in soil and manure. After the bacteria enter an injury, such as a cut or puncture wound, they produce a toxin which is absorbed and causes the disease. The disease affects all animals, with the horse being particularly susceptible. Tetanus is characterized by stiffness of any or all muscles. Advanced symptoms include violent spasms, stiff tail and legs, sweating, and fever. Once symptoms occur the disease is usually fatal. Vaccination for all horses is recommended. Initially, the horse should be vaccinated with tetanus toxoid twice, three to four weeks apart, then revaccinated annually and following any cut. If the horse has not been previously vaccinated with tetanus toxoid or vaccination history is unknown, tetanus antitoxin should be given following any cut and a tetanus toxoid vaccination program begun two to four weeks later.

Tetany: A condition in an animal in which there are localized, spasmodic muscular contractions, twitching, or cramps.

Therm: An expression of nutritional energy. One therm equals 1000 nutritional *calories* or *kilocalories* (kcal), and 1 *megacalorie* (Mcal).

Thiabendazole: A drug that is effective in killing many types of intestinal worms. It is also effective against some fungi.

Thrush: A bacterial infection of the sulci of the frog (see Glossary Fig. 1) characterized by the presence of a black *necrotic* material with a very offensive odor. The infection may penetrate the horny tissue and involve the sensitive structures, resulting in lameness. The major cause is standing in dirty, wet stalls. It is treated by eliminating the cause, and removing degenerated frog tissue. The foot should be cleaned, and the sulci of the frog packed with cotton soaked in sulfapyridine daily until the infection is controlled.

Thumps: See *synchronous diaphragmatic flutter.*

Tibia: The major bone between the *stifle* and the *hock* (see Glossary Fig. 1).

Total Digestible Nutrients (TDN): A term which indicates the *energy* density of a feedstuff. It is the sum of the percent of digestible *fat* times 2.25, digestible *protein*, and digestible *carbohydrate* (NFE and *fiber*) present in the feed. One pound of TDN is equal to approximately 2000 *kcalories* of digestible energy.

Toxoid: see *vaccine.*

Trace mineral: A *mineral* required by animals in their *diet* in very small amounts (milligrams per pound of diet or less). See minerals.

Transferrin: The *protein* which transports iron in the blood.

Tying-up syndrome: Muscle spasms that occur after a period of prolonged or hard physical activity due to a depletion of the *energy* necessary for muscle relaxation. Not to be confused with *azoturia* (see Table 5–3).

U

Umbilical stump: The 1 to 2 inch (2 to 5 cm) stump on the newborn's abdomen that, prior to birth, was attached to the *placenta*. Most of the young animal's nourishment prior to birth is passed to it from the mother through the umbilicus. Following birth, bacteria may gain access to the body via the umbilical stump. These bacteria may be carried by the blood and localize in the lungs, resulting in pneumonia, the brain, frequently resulting in death, or most commonly, in the joints, resulting in infected, swollen joints, and lameness. To prevent this, the umbilical stump should be soaked for 1 to 2 minutes in 7% tincture of iodine as soon as possible following birth.

Urea: Consists chemically of two molecules of ammonia attached to one molecule of carbon monoxide (NH_2-CO-NH_2). It is produced in the liver and excreted primarily in the urine, and is the major means of the body's excreting nitrogen. Nitrogen is produced from the breakdown of body or dietary *protein*. Some urea produced by *ruminants* goes into the *rumen* (urea may also be fed to ruminants). In the rumen, urea is broken down to carbon dioxide and ammonia. Bacteria in the rumen utilize the ammonia to produce protein. The animal is able to digest and absorb this protein as it passes through the digestive tract (see Chapter 1).

Urticaria: See *hives* and *protein bumps*.

Uterine: Pertaining to the uterus which is a hollow, muscular organ in female animals where the young develop and grow prior to birth.

V

Vaccinate (Inoculate): The administration of a vaccine to an animal. Depending on the vaccine, it may be given in a number of different ways, such as orally, sprayed into the nose, or injected into the muscle, vein, skin, or under the skin (subcutaneously).

Vaccine: A preparation containing a virus, bacteria, or other infectious organism or toxin given to an animal to stimulate it to produce an immunity against that organism or toxin, and therefore, prevent the disease caused by that organism or toxin. When the vaccine contains a toxin, it is called a toxoid. The toxoid stimulates the animal to produce antibodies, called antitoxins, against that toxin.

Vein: A thin walled tube that carries blood from all parts of the body to the heart. The blood in a vein is normally low in oxygen because it has diffused out of the blood into the tissue. This gives it a dark red color, whereas the blood in an *artery* is normally bright red because it is saturated with oxygen which it has picked up in its passage through the lungs. Veins are much closer to the body surface than are arteries. The blood in a vein is under little pressure, so that if a vein is cut, blood does not spurt out as it does when an artery is cut.

Vertebrae: Bones making up the spinal column or back bone.

Vitamins: Organic (carbon-containing) compounds needed in minute amounts for normal body function, but which do not supply *energy* and are not a part of body structure. Some are produced in the horse's body (vitamins C and D), some are produced by bacteria in the intestinal tract (all of the B-vitamins and vitamin K), and some are supplied entirely by the *diet* (vitamins A and E).

W

Water bag: See *amnionic sac.*

Wind sucker: A mare that, particularly when running, aspirates air, and usually also fecal material, into the vagina. This results in inflammation of the vagina and sometimes of the *uterus.* It is corrected by surgically suturing the upper portion of the lips of the vulva together. This is called a Caslick's operation. A horse that *cribs* or chews wood is sometimes incorrectly referred to as a wind sucker.

Wobbler (Wobbler's syndrome): Incoordination of the back legs generally occurring in the growing horse, and resulting from damage to the spinal cord of the neck which may be caused by injury, or nutritionally-induced *osteoarthritis* or *osteochondritis* of the *vertebrae.* The nutritional inducements most commonly incriminated are feeding of excess *concentrates,* or inadequate calcium or phosphorus.

Wood chewing: Chewing of wood by the horse, generally as a result of boredom, from habit, or because the horse likes its taste. Not known to be due to a nutritional deficiency or imbalance. Most horses do not swallow the wood. Should not be confused with *cribbing* or *wind sucking* (see Chapter 4).

Worms: Refers to worms and botfly larvae in the horse's stomach and intestines. See *ascarids, bots, strongyles,* and *pinworms.* Worming an animal means treating the animal to remove these worms (see discussion of intestinal parasites in Chapter 10).

X

X-rays: See *radiographs.*

Appendix A

APPENDIX TABLE 1
NUTRIENT CONTENT OF COMMONLY USED FEEDSTUFFS[50]*

	Digestible Energy		TDN %	Crude Fiber %	Crude Protein %	Cal-cium %	Phos-phorus %
	(Mcal/kg)	(kcal/lb)					
ALFALFA							
grazed or hay, prebloom	2.25	1025	52	20	19		
hay—early bloom	2.15	980	50	28	17	1.0†	0.22†
hay—mid-bloom and	2.1	940	47	30	14	(0.6–	(0.10–
grazed full-bloom						2.0)	0.30)
hay—full-bloom	2.0	900	44	31	13		
dehy, 17%	2.25	1020	50	24	17		
dehy, 15%	2.15	980	49	30	15		
BAHIAGRASS							
grazed	1.9	855	43	29	7	0.40	0.2
hay	1.7	780	39	27	5		
BARLEY							
grain	3.25	1470	74	5	12	0.04	0.30
grain, Pacific Coast	3.15	1430	71	6	10	0.04	0.30
hay	1.7	780	40	24	8	0.2	0.25
straw	1.45	650	33	38	4	0.2	0.04
BEET							
pulp, dehydrated	2.5	1140	58	20	7	0.6	0.1
BERMUDA GRASS							
grazed	2.0	900	45	25	8	0.4	0.25
hay	1.8	820	41	31	6	0.35	0.25
BIRDSFOOT TREFOIL	2.0	900	45	27	14	1.5	0.2

221

APPENDIX TABLE 1 (Continued)
NUTRIENT CONTENT OF COMMONLY USED FEEDSTUFFS[50]*

	Digestible Energy		TDN %	Crude Fiber %	Crude Protein %	Cal-cium %	Phos-phorus %
	(Mcal/kg)	(kcal/lb)					
BLUEGRASS, KENTUCKY							
grazed, early	2.15	980	50	23	15	0.4	0.35
grazed, posthead	2.0	900	45	24	10	0.35	0.35
hay	2.0	900	45	27	10	0.3	0.3
BRAN (see wheat)							
BREWERS							
grains, dehydrated (Coor's pellets, etc.)	2.7	1230	61	14	25	0.3	0.5
BROME							
grazed, vegetative	2.7	1230	61	22	17	0.4	0.3
hay, late bloom	2.15	980	49	36	6	0.3	0.2
CALF MANNA‡	—	—	—	6	25	1.0	0.6
CANARYGRASS, REED							
grazed	2.15	980	49	26	11	0.35	0.3
hay	1.9	855	44	30	11	0.35	0.2
CITRUS PULP							
w/o fines, dehydrated	2.7	1230	61	13	6	1.9	0.1
CLOVER, ALSIKE							
hay	1.9	855	43	26	14	1.2	0.3
CLOVER, CRIMSON							
grazed	2.15	980	50	24	15		
hay	1.9	855	44	29	16		
CLOVER, LADINO							
hay	2.0	900	46	18	19	1.2† (0.6–2.0)	0.22† (0.10–0.30)
CLOVER, RED							
grazed, early bloom	2.25	1025	52	17	19		
grazed, late bloom	2.15	980	50	27	13		
hay	1.9	855	44	27	13		

CORN							
cobs, ground	1.25	570	28	32	3	0.1	0.04
distillers grains, dehy	2.8	1270	63	11	27	0.1	0.40
ears, ground	3.0	1350	67	9	8	0.02	0.25
grain	3.5	1595	80	2	10	0.02	0.25
COTTONSEED							
meal	2.4	1070	68	12	41	0.15	0.65
whole	3.7	1750	89	18	22	0.15	0.65
HORSE CHARGES	—	—	—	7.5	33	1.5	0.9
FESCUE, MEADOW							
grazed	2.1	940	47	26	10	0.5	0.35
hay	1.8	820	41	30	9	0.5	0.30
LESPEDEZA							
grazed, mature	2.0	900	45	34	14	1.0	0.20
hay	1.8	820	42	29	13	1.0	0.30
LINSEED MEAL	2.7	1230	62	9	34	0.35	0.8
MILK, skimmed, dehydrated	3.7	1655	83	0.3	32	1.2	1.0
MOLASSES							
beet, wet	2.9	1310	65	—	8	0.2	0.03
sugarcane, dehydrated	2.9	1310	65	4.5	8	0.7	0.20
sugarcane, wet	2.9	1310	66	—	4	0.9	0.15
OATS							
grain	3.0	1350	68	11	12	0.1	0.3
grain, Pacific Coast	3.0	1350	69	11	9	0.1	0.3
hay	1.8	820	42	29	8	0.25	0.20
straw	1.9	700	35	36	4	0.2	0.04
ORCHARDGRASS							
grazed	2.2	980	50	24	16	0.4	0.45
hay	1.8	820	42	32	9	0.25	0.25
PANGOLAGRASS							
grazed	2.0	900	46	26	11	0.35	0.25
hay	1.8	820	40	32	8	0.25	0.20

APPENDIX TABLE 1 (Continued)
NUTRIENT CONTENT OF COMMONLY USED FEEDSTUFFS[50]*

	Digestible Energy		TDN %	Crude Fiber %	Crude Protein %	Cal-cium %	Phos-phorus %
	(Mcal/kg)	(kcal/lb)					
PRAIRIE							
hay	1.8	820	41	30	5	0.35	0.15
RYE							
grain	3.2	1430	72	3	12	0.04	0.3
START TO FINISH‖	4.0	1820	91	1.5	30	1.5	1.0
SORGHUM							
grain (milo)	3.2	1430	72	3	11	0.03	0.3
SOYBEAN							
hay	1.9	855	43	31	14	1.1	0.20
hulls	2.3	1060	54	36	11	0.35	0.15
seeds	3.6	1640	83	5	40	0.25	0.5
meal	3.2	1470	74	6	44	0.25	0.6
SUNFLOWER MEAL	2.8	1270	64	11	45	0.35	1.0
TIZWHIZ 30-PLUS#	3.2	1470	74	6	30	1.5	1.1

TIMOTHY							
grazed, mid-bloom	1.9	855	44	28	8	0.25	0.2
hay, s-c, pre-head	2.0	900	45	28	8	0.4	0.2
hay, s-c, head	1.8	820	40	29	6	0.35	0.15
TREFOIL, BIRDSFOOT							
hay	2.0	900	45	27	14	1.5	0.2
WHEAT							
bran	2.8	1190	60	10	15	0.1	1.2
grain	3.4	1550	78	3	12	0.05	0.35
hay	1.7	780	39	26	7	0.13	0.15
straw	1.3	610	31	37	4	0.2	0.04
YEAST							
brewer's, dehydrated	3.0	1350	68	3	43	0.4	1.3

*Values given are those present in air dried feed. Air dry feeds contain 10% moisture and 90% dry matter. This is the approximate moisture content of most feeds in the form in which they are fed to the horse. Feeds not given in this table may be found in the National Research Council Atlas of Nutritional Data on U.S. and Canadian Feeds, available from the National Academy of Sciences, Printing and Publishing Office, 2101 Constitution Avenue, Washington, D.C. 20418.

†Calcium and phosphorus content are not greatly influenced by the stage of maturity of the feeds given; therefore, the most frequent amount generally present in the feed is given, followed by the range, given in parentheses.

‡Carnation-Albers, 6400 Glenwood, Box 2917, Shawnee Mission, KS 66201.

§Ralston Purina Co., Checkerboard Square, St. Louis, MO 63188.

‖Milk Specialties Co., Box 278, Dundee, IL 60118.

#Tizwhiz Distributors Inc., P.O. Box 604, Worthington, OH 43085.

MINIMUM NUTRIENT REQUIREMENTS[50]

	Energy		Required in Total Air Dry Ration (%)			Percent of Body wt. eaten/day†
	Mcal/ day*	Lb TDN /day*	Crude Protein	Calcium	Phosphorus	
Mature horse at rest‡ and first 8 mo of pregnancy	16.4	8.2	8.0	0.30	0.20	1.5
Last 3 mo of pregnancy	18.4	9.2	10	0.45	0.35	1.75
Lactation, first 3 mo	28.3	14.1	12.5	0.45	0.35	2.75
Lactation, after 3 mo	24.3	12.2	11	0.40	0.30	2.25
2 yr old to maturity	16.5	8.2	9	0.40	0.35	1.75
18 to 24 mo	17.0	8.6	10	0.40	0.35	2.0
12 to 18 mo	16.8	8.4	12	0.50	0.35	2.5
Weanling	15.6	7.8	14.5	0.65	0.45	3.0
Nursing foal, 3 to 5 mo, requirements above milk	6.9	3.5	16	0.80	0.55	0.75§

*Amount needed by the 1100 lb horse (500 kg). For each 100 lbs (45 kg) above or below 1100 lbs body weight (500 kg), add or subtract 8% to this amount. To convert lb TDN/day to kg TDN/day, divide by 2.2.

†The approximate amount of air dry feed containing 50% TDN that the horse will eat daily.

‡See Appendix Table 3 for the energy needs for physical activity. All other nutritional needs for physical activity are the same as they are for rest. The maximum amount that the horse's stomach and intestinal tract will hold daily is equal to about 3% of the horse's body weight.

§The maximum amount of creep ration consisting entirely of concentrates that should be fed daily.

APPENDIX TABLE 3
ENERGY REQUIREMENTS FOR PHYSICAL ACTIVITY[50]
(ABOVE THAT NEEDED AT REST)*

Physical Activity	Mcal/Hr/100 lbs (45 kg) Body wt	Lbs TDN/hr/ 100 lbs Body wt†
Walking	0.02	0.01
Slow trot	0.23	0.12
Fast trot and cantering	0.57	0.29
Cantering and galloping	1.05	0.53
Strenuous effort	1.77	0.89

*Example: An 1100 lb horse at a slow trot for 3 hrs would need (1100 lbs body wt) × (0.12 lbs TDN/hr/100 lbs body wt) × (3 hrs) = 4.0 lbs of TDN for physical activity plus 8.2 lbs of TDN at rest (from Appendix Table 2) for a total of 12.2 lbs of TDN per day. The lbs of feed necessary to provide this amount of TDN would be 12.2 ÷ the % TDN in the feed used (Appendix Table 1), e.g., 24.4 lbs of early bloom alfalfa hay (12.2 ÷ 0.5 = 24.4 lbs).

†Or kgs TDN/hr/100 kgs body wt.

APPENDIX TABLE 4
DETERMINING THE HORSE'S WEIGHT

Girth Length inches (cm)		Weight lbs (kgs)	
30	(76)	100	(45.5)
40	(102)	200	(91)
45.5	(116)	300	(136.5)
50.5	(128)	400	(182)
55	(140)	500	(227)
58.5	(148)	600	(273)
61.5	(156)	700	(318)
64.5	(164)	800	(364)
67.5	(171)	900	(409)
70.5	(178)	1000	(455)
73	(185)	1100	(500)
75.5	(192)	1200	(545)
77.5	(197)	1300	(591)

APPENDIX TABLE 5
CALCIUM AND PHOSPHORUS CONTENT OF MINERAL SUPPLEMENTS*

Mineral Supplements	% Calcium	% Phosphorus
Biophos	18	21
Bone Meal	24–32	12–14
Dicalcium phosphate (Dical)	20–27	18–21
Limestone (calcium carbonate)	33–36	0
Monosodium or disodium phosphate (monophos, XP-4)	0	22–27
Monodicalcium phosphate	15–21	22
Rock Phosphate, defluorinated	29–36	12–18
Sodium Tripolyphosphate (polyphosphate, XP-4)	0	25
Calcite	34	0
Oyster shells	35–38	0
Gypsum (calcium sulfate = 18% sulfur)	22	0
Phosphoric Acid	0	24
Diammonium phosphate	0	20–23
Rock Phosphate, soft	18	9
Purina's 12:12 (+ 5% TM salt)†	12	12
Dairy Phos Mineral†	0	16
CO-OP Perfect 36 (+ 12% TM salt)†	12	12
CO-OP Hi Ratio (+ 30% TM salt)†	16	4
CO-OP Pro-Ten − 4 (+ 20% TM salt)†	8	4
CO-OP Dairy Phos (+ 8% TM salt)†	16	15
CO-OP OP-T-MIN (+ 8% TM salt)†	8	16

*Many of these mineral supplements are available at most feedstores but may be called by names other than those given here.

†The CO-OP (Farmland Industries, Inc., Kansas City, MO 64116) and Purina (Ralston Purina Co., Checkerboard Square, St. Louis, MO 63188) minerals contain 100,000 IU of vitamin A per pound, and Purina 12:12 contains cottonseed meal and molasses to increase its palatability and consumption. There are many commercial calcium and phosphorus supplements available. These are given as examples. Their inclusion here does not indicate that they are any better or worse, or are recommended, as compared with other commercial supplements.

APPENDIX TABLE 6
FEEDING PROGRAMS AND CONCENTRATE MIX EXAMPLES

GRAIN MIX INGREDIENTS

Percent of ingredients in the grain mix*

Mix No.	Any Cereal grains	Soybean meal† (44% protein)	Cottonseed meal† (41% protein)	Wet Molasses‡	Dicalcium Phosphate	Limestone
1	50	18.75	17.5	10	2.5	1.25
2	62.5	12.5	12.5	10	1.25	1.25
3	80	12.5	0	5	2.5§	0
4	93.5	0	0	5	1.5§	0

*After a grain mix is prepared, it is recommended that it be analyzed to see that it contains the nutrients given below. Mixing errors and poor mixing, unfortunately, are not uncommon. If, instead of having the ration mixed, the individual ingredients are fed, 1% mineral is approximately 1 tsp, or 0.2 oz/lb of feed, and 1% is 0.01 lbs/lb of feed. Example: if 2.5% dical and 12.5% soybean meal are required, and 6 lbs of grain/day is being fed, feed $(2.5) \times (0.2 \text{ oz.}) \times (6) = 3$ oz of dical, and $(1.25) \times (0.01) \times (6) = 0.75$ lbs of soybean meal and 5.25 lbs $(6 - 0.75)$ of grain daily.

†Protein and mineral supplements will frequently sift out when the total of the two makes up greater than 20 to 25% of the concentrate mix. If adding additional molasses to the concentrate mix does not prevent this, the concentrate mix should be pelleted. Soybean meal is the protein supplement preferred for the growing horse because of its high lysine content. However, to increase the palatability of a concentrate mix that contains more than 20% protein supplement, as do mixes numbers 1 and 2, it is best to use one-half soybean meal and one-half cottonseed or linseed meal, instead of using all soybean meal.

‡The amount of molasses can be varied as necessary to hold the ration together and decrease dustiness, yet not make the ration stick together in compact clumps. More than 10 to 15% molasses may cause the mix to clump. In hot, humid areas concentrate mixes containing greater than 5% molasses may become moldy. If less than 10% molasses is used in mix numbers 1 and 2, pelleting may be necessary to prevent the protein and mineral supplements from sifting out.

§Monosodium phosphate may be substituted for dicalcium phosphate in these rations, but only when they are fed with alfalfa.

NUTRIENTS IN GRAIN MIXES AND TOTAL RATION*

Mix	Nutrients in grain mix (%)†			Nutrients in total ration (%)‡		
	Protein	Ca	P	Protein	Ca	P
No. 1	22	1.1	0.85	14.5	0.70	0.50
No. 2	17	0.75	0.55	12.5	0.55	0.35
No. 3	13.5	0.6	0.7	14.5	0.85	0.45
No. 4	11	0.4	0.55	12.5	0.70	0.35
Super Foal§	16	1.3	0.8			
Big Un#	16	0.6	0.6	(8% crude fiber)		
Tizwhiz||	16	0.8	0.6	(6% crude fiber & 83% TDN or 1660 kcal/lb)		
Tizwhiz–12||	12	0.8	0.6			
Omolene or Pure Pride 300#	16	0.6	0.6			
Omolene or Pure Pride 200#	14	0.45	0.45			
Omolene or Pure Pride 100#	10	0.45	0.45			
Omolene Checkers#	13	0.45	0.3			
Super Horse#	14	1.0–0.5	0.5			
Trophy§	12	1.0–0.5	0.5			
Pride of Arena**						
Textured	12	—	—			
Fourteen	14	—	—			
CO-OP Block	28	1.5–2.5	1.0	(33½ lb block, 9.5% crude fiber)		
Horse Block#	12.5	1.7–0.8	0.6	(33½ lb block, 12% crude fiber)		
Sweetena#,††	10	(20% roughage, 15% crude fiber & 65% TDN or 1300 kcal/lb)				
Horse & Mule Chow#,††	10	(20% roughage, 15% crude fiber & 65% TDN or 1300 kcal/lb)				
Horse Feed Complete**,††	13	(60 to 70% roughage, 22% crude fiber & 59% TDN or 1180 kcal/lb)				
Horse & Chow Checkers#,††	12.5	(70 to 80% roughage, 25% crude fiber & 57% TDN or 1140 kcal/lb)				
Roundup§,††	12	(70 to 80% roughage, 25% crude fiber & 57% TDN or 1140 kcal/lb)				

*The commercial rations given are intended as examples of common ones available. Their inclusion here doesn't indicate that they are any better or worse, or are recommended, as compared with other commercial rations.

†Nutrient contents in mixes numbers 1 through 4 are based on using air dry feeds containing the following percent protein, calcium, and phosphorus respectively: grain 12, 0.05 and 0.03; soybean meal 44, 0.3 and 0.65; dicalcium phosphate 0, 24 and 19; and limestone 0, 33 and 0. All commercial mixes contain 70 to 74% TDN (1400 to 1480 kcal/lb) and less than 10% crude fiber, unless stated otherwise.

‡See following for feeding program recommended for that grain mix which results in the nutrient contents given for the total ration, i.e., grain mix plus roughage.

§Carnation—Albers, 6400 Glenwood, Box 2917, Shawnee Mission, KS 62201.

"Tizwhiz Distributors Inc., P.O. Box 604, 657 High Street, Worthington, OH 43085.

#Ralston Purina Co., Checkerboard Square, St. Louis, MO 63188.

**CO-OP, Farmland Industries, Inc., Kansas City, MO 64116.

††Rations containing greater than 20% crude fiber may be fed as the sole ration for the horse. All others are to be fed with roughage.

APPENDIX TABLE 6
FEEDING PROGRAMS AND CONCENTRATE MIX EXAMPLES (Continued)

FEEDING PROGRAMS

Horse	Roughage* Type	Grain Mix Mix No.	Grain Mix Maximum (lbs/100 lbs/day) or (kgs/100 kgs/day)	Salt-Mineral† mix always available
Creep feeding (nursing foal)	Either	2	0.75	TM salt
Weanling (6–12 mo)	Legume	3	1.5	adequate, but
Yearling (1–2 yrs)	Legume	4	1.0	High-P okay
Lactating mare	Legume	4	as needed	High-P
Last 3 mo pregnancy	Legume	4	as needed	High-P
Weanling (6–12 mo)	Grass	1	1.5	TM salt adequate, but balanced Ca
Yearling (1–2 yrs)	Grass	2	1.0	and P okay
Lactating mare	Grass	2	0.5–0.7	Balanced Ca & P
Last 3 mo pregnancy	Grass	3	0–0.5	Balanced Ca & P
All others	Either	Any cereal grain as needed		TM salt

*The nutrients present in the total air dry ration as given in the preceding table are based on average or better quality air dry legumes (alfalfa, trefoil, clover, or lespedeza) containing 17% protein, 1.0% calcium, and 0.20% phosphorus, and air dry grass hay containing 7% protein, 0.35% calcium, and 0.20% phosphorus. If the nutrients present in the roughage are lower than this, a grain mix higher in nutrients than the one recommended should be fed. Regardless of the nutrient contents of the roughage, a grain mix higher in nutrients can always be substituted for one lower in nutrients, but not the reverse.

†High-P mineral mix is one containing trace mineralized salt (TM salt), 6 to 10% calcium (Ca) and 14 to 18% phosphorus (P), such as: equal parts of TM salt, calcium phosphate, and sodium phosphate; or one such as CO-OP's OP-T-MIN (Farmland Industries, Inc., Kansas City, MO 64116). A balanced calcium and phosphorus mineral mix is one containing TM salt, 8–14% Ca and 8–14% P, such as: 1/2 TM salt + 1/2 calcium phosphate; or Purina's 12:12; or CO-OP's Perfect 36 (see Appendix Table 5).

APPENDIX TABLE 7
GROWTH RATE OF THE LIGHT HORSE*

	% of Mature	
Age (months)	Weight	Height at Withers
birth	8–9	61–64
1	16–18	66–68
3	27–29	75–77
6	45–47	83–86
9	56–58	89–91
12	65–69	91–93
18	78–83	94–96
24	87–92	96–98
30	93–97	97–99
36	95–99	98–100
48	98–100	99–100

*The larger horse (1300 lbs or 590 kgs) will be on the low end, and the smaller horse (700 lbs or 318 kgs), on the upper end of the range given (adapted from [50] and [56]). Draft horses' rate of growth is slower than the values given, e.g., 50% of mature weight is attained at 1 year, 75% at 2, and 90% at 3 years of age;[9] and ponies growth rate is faster (55% is attained at 6 months, 75% at 1 year and 84% at 18 months of age).[103] Body weight, height at withers, and circumference of front cannon bones are lowly heritable at birth but are highly heritable by 18 months of age (0.90, 0.88, and 0.77, respectively).[103] Colts are heavier and taller at birth and have bigger cannon bones than fillies, and the differences increase with time.[103] All three of these parameters are smaller in foals born to mares less than 7 or older than 12 years old than in foals born to mares between these ages.[103]

APPENDIX TABLE 8
CONVERSION FACTORS*

Units Given	Units Wanted	For Conversion Multiply by	Units Given	Units Wanted	For Conversion Multiply by
Area:					
acres	hectare	0.4047	lbs/bu	kg/kl	12.9
acres	sq ft	43,560	lbs barley	bu	48
acres	sq rods	160	lbs ear corn	bu	70
acres	sq yds	4,840	lbs gd wheat	bu	50
hectare	sq meters	1,000	lbs gd oats	bu	23
section	acres	640	lbs shelled corn, milo or rye	bu	56
sections	sq miles	1			
sq ft	sq meters	0.0929			
sq inches	sq cm	6.452	lbs whole oats	bu	32
sq miles	sq km	2.6			
sq yds	sq meters	0.836	lbs whole wheat or soybeans	bu	60
township	section	36			
Concentration:					
g/kg	%	0.1	qts bran, alfalfa meal, dried beet pulp	lbs	0.5–0.6
lbs/short ton	kg/metric ton	0.5			
mg/g	mg/lb	453.6			
mg/g	%	0.1	qts most grains and soybean meal	lbs	1.5–1.9
mg/kg	mg/lb	0.4536			
mg/kg or l	%	0.0001			
mg/lb	g/short ton	2	qts oats, linseed meal	lbs	1.0
%	g/short ton	9.072	ton baled hay	cu ft	200–360
%	lbs/gal	8.34	ton loose hay	cu ft	450–600
%	mg %	1,000	ton cubed hay	cu ft	60–70
%	oz/gal	1.28	ton baled straw	cu ft	400–500
ppm	g/ton	0.9072	ton loose straw	cu ft	670–1,000
ppm	mg/g	0.001			
ppm	mg/kg or l	1	**Energy:**		
ppm	μg/g or ml	1	Calorie	nutritional calorie	1
ppm	mg/lb	0.4536			
ppm	%	0.0001	kcal	Calorie	1
ppm	ppb	1,000	kcal/kg	kcal/lb	0.4536
μg/kg	μg/lb	0.4536	kcal/lb	kcal/kg	0.2046
			lbs TDN	kcal	2,000
Density:			Mcal	kcal	1,000
cu ft	bu corn	0.8	Therm	Mcal	1
cu ft	bu ear corn	0.4			

APPENDIX TABLE 8
CONVERSION FACTORS* (Continued)

Units Given	Units Wanted	For Conversion Multiply by	Units Given	Units Wanted	For Conversion Multiply by
Length:			**Volume:**		
fathoms	feet	6.08	barrels	gallons	31.5
feet	centimeters	30.48	bushels	cubic feet	1.25
furlongs	yard	220	bushels	gallons	9.31
hands	inches	4	bushels	hectoliters	0.3524
inches	millimeters	25.4	bushels	pecks	4
leagues	miles	3	cubic feet	gallons (water)	7.48
miles	furlongs	8	cubic inch	ml or cc	16.387
miles	feet	5,280	cup	oz	8
miles	kilometers	1. 609	cord	cu ft	128
nautical miles	land miles	1.15	gallon	cu inch	231
rods	feet	16.5	gallon	lbs water	8.35
yards	meters	0.914	gallon	qt	4
Metric Unit Prefixes:			hogshead	barrel	2
M-unit	unit	1,000,000	liter	gal	0.2642
kilo-unit	unit	1,000	liter	qt	1.057
hecto-unit	unit	100	ml	oz	0.034
deca-unit	unit	10	oz	ml	29.57
deci-unit	unit	0.1	peck	qt	8
centi-unit	unit	0.01	pint	ml	473
milli-unit	unit	0.0001	pint	oz	16
micro-unit	unit	0.000001	qt	liter	0.946
Miscellaneous:			qt	pt	2
mg beta-carotene	IU of vitamin A for horses	400	Tbsp	tsp	~3
nitrate nitrogen (NO_3-N)	nitrate	4.4	tsp	g or ml	~5
			Weight:		
potassium nitrate (KNO_3)	nitrate (NO_3)	0.6	dram	g	1.77
			grain	mg	64.8
			lb	g	453.6
sodium nitrate ($NaNO_3$)	nitrate	0.7	lb	kg	0.4536
			long ton	lbs	2,200
Temperature:			long ton	metric ton	1
°F	°C	5/9 after subtracting 32	oz	g	28.35
°C	°F	9/5 then add 32	short ton	lbs	2,000
			short ton	metric ton	0.9072

*To make the opposite conversion, divide by the conversion number given in the table instead of multiplying by it. For example, 5 lbs = 5 × 453.6 = 2268 g, or 2268 g ÷ 453.6 = 5 lbs.

APPENDIX TABLE 9
BREED ASSOCIATIONS AND REGISTRIES
(listed in alphabetical order by breed)

Association or Registry	Address	City and State
Creme & White Horse Registry (Albino)	P.O. Box 79	Crabtree, OR 97335
Andalusian Horse Registry of America	Box 1290	Silver City, NM 88061
The Appaloosa Horse Club, Inc.	P.O. Box 8403	Moscow, ID 83843
National Appaloosa Pony, Inc.	P.O. Box 206	Gaston, IN 47342
Arabian Horse Registry of America	3435 South Yosemite	Denver, CO 80231
International Arabian Horse Assoc.	P.O. Box 4502	Burbank, CA 91503
American Bashkir Curly Registry	Box 453	Ely, NV 89301
American Bay Horse Assoc.	P.O. Box 884F	Wheeling, IL 60090
Belgian Draft Horse Corp. of America	P.O. Box 335	Wabash, IN 46992
American Buckskin Registry Assoc.	P.O. Box 1125	Anderson, CA 96007
International Buckskin Horse Assoc.	P.O. Box 357	St. John, IN 46373
The Chickasaw Horse Assoc. Inc.	Box 607	Love Valley, NC 28677
Cleveland Bay Society of America	P.O. Box 182	Hopewell, NJ 08525
Clydesdale Breeders of the U.S.	Route 1, Box 131	Pecatonica, IL 61063
American Connemara Pony Society	R.D. 1	Goshen, CT 06756
American Crossbred Pony Registry	Box M	Andover, NJ 07821
National Cutting Horse Assoc.	P.O. Box 12155	Ft. Worth, TX 76116
The American Donkey & Mule Society, Inc.	Route 5, Box 65	Denton, TX 76201
Endurance Horse Registry of America	P.O. Box 63	Agoura, CA 91301
American Fox Trotting Horse Breed Assoc.	Box 666	Marshfield, MO 65706
Galiceno Horse Breeders Assoc.	111 East Elm Street	Tyler, TX 75702
American Gotland Horse Assoc.	RR. #2, Box 181	Elkland, MO 65644
American Hackney Horse Society	P.O. Box 174	Pittsfield, IL 62363
Haflinger Assoc. of America	2624 Bexley Park Road	Bexley, OH 43209
Half-Quarter Horse Registry	43949 North 60th West	Lancaster, CA 93534
The Half Saddlebred Registry of America	660 Poplar Street	Coshocton, OH 43812
Half-Thoroughbred American Remount Assoc.	P.O. Box 1066	Perris, CA 92370
American Hanoverian Society	809 West 106th Street	Carmel, IN 46032
American Holstein Horse Assoc.	12 Southlawn Avenue	Dobbs Ferry, NY 10522
Horse of the Americas Registry	248 North Main Street	Porterville, CA 93257
Hungarian Horse Assoc.	Bitteroot Stock Farm	Hamilton, MT 59840
The Hunter Club of America	Box 274	Washington, MI 48094
American Hunter & Jumper Assoc.	P.O. Box 1174	Fort Wayne, IN 46801
Icelandic Pony & Registry	56 Alles Acres	Greeley, CO 80631
American Indian Horse Registry	Route 1, Box 64	Lockart, TX 78644
Standard Jack & Jennet Registry	300 Todds Road	Lexington, KY 40511
Lipizzan Assoc. of America	Woolworth Tower	New York, NY 10279
Royal International Lippizzaner Club of America	Route 7	Columbia, TN 38401
Miniature Horse Registry (International)	P.O. Box 907	Palos Verdes Estates, CA 90274
Miniature Donkey Registry of U.S. Inc.	1108 Jackson Street	Omaha, NE 68102
Missouri Fox Trotting Horse Breed	P.O. Box 637	Ava, MO 65608
Morab Horse Registry of America	P.O. Box 143	Clovis, CA 93613
American Morgan Horse Assoc.	P.O. Box 1	Westmoreland, NY 13490
American Mustang Assoc., Inc.	P.O. Box 338	Yucaipa, CA 92399
National Mustang Assoc., Inc.		Newcastle, UT 84756
The Spanish Mustang Registry	Route 2, Box 74	Marshall, TX 75670

APPENDIX TABLE 9
BREED ASSOCIATIONS AND REGISTRIES (Continued)
(Listed in alphabetical order by breed)

Association or Registry	Address	City and State
Norwegian Fjord Horse Assoc., of North America	RR. 1 Box 370	Round Lake, IL 60073
American Paint Horse Assoc.	P.O. Box 18519	Forth Worth, TX 76118
Palomino Horse Assoc., Inc.	P.O. Box 324	Jefferson City, MO 65102
Palomino Horse Breeders of America	P.O. Box 249	Mineral Wells, TX 76067
American Part-Blooded Horse Registry	4120 S.E. River Drive	Portland, OR 97222
Part-Thoroughbred American Remount Assoc.	11783 N. Ranch Lane	Scottsdale, AZ 85260
Paso Fino Owners & Breeders Assoc.	P.O. Box 764	Columbus, NC 28722
Percheron Horse Assoc. of America	Route 1	Belmont, OH 43718
Peruvian Paso Half-Blood Assoc.	323–33rd Road	Palisade, CO 81526
American Peruvian Paso Horse Registry	Route 3, Box 331B	Boerne, TX 78006
Peruvian Paso Horse Registry of North America	P.O. Box 816	Guerneville, CA 95446
American Assoc. of Owners' & Breeders' of Peruvian Paso Horses	P.O. Box 2035	California City, CA 93505
Pinto Horse Assoc. of America	7525 Mission Gorge Road, Suite C	San Diego, CA 92120
Pony of the Americas Club, Inc.	P.O. Box 1447	Mason City, IA 50401
American Quarter Horse Assoc.	2736 West Tenth Street	Amarillo, TX 79168
National Quarter Horse Registry	Box 235	Raywood, TX 77582
Standard Quarter Horse Assoc.	4390 Fenton Street	Denver, CO 80212
National Quarter Pony Assoc., Inc.	Route 1, Box 585	Marengo, OH 43334
Racking Horse Breeders Assoc.		Helena, AL 35080
Colorado Ranger Horse Assoc.	7023 Eden Mill Road	Woodbine, MD 21797
Ysabelle Saddle Horse Assoc.	R.R. #2	Williamsport, IN 47993
American Saddlebred Horse Assoc.	929 South Fourth Street	Louisville, KY 40203
American Shetland Pony Club	P.O. Box 435	Fowler, IN 47944
American Shire Horse Assoc.	14410 High Bridge Road	Monroe, WA 98272
Spanish-Barb Breeders Assoc.	P.O. Box 7479	Colorado Springs, CO 80907
The Spanish Mustang Registry	Route 4, Box 64	Council Bluffs, IA 51501
American Council of Spotted Asses	2126 Fairview Place	Billings, MT 59102
U.S. Trotting Assoc. (Standardbred)	750 Michigan Avenue	Columbus, OH 43215
American Suffolk Horse Assoc.	15B Roden	Wichita Falls, TX 76311
American Tarpan Studbook Assoc.	Route 6, Box 429	Griffin, GA 30223
Tennessee Walking Horse Breeders Assoc.	P.O. Box 286	Lewisburg, TN 37091
The Jockey Club (Thoroughbred)	380 Madison Ave.	New York, NY 10017
American Trakehner Assoc.	P.O. Box 132	Brentwood, NY 11717
North American Trakehner Assoc.	Box 100	Bath, OH 44210
International Trotting & Pacing Assoc.	575 Broadway	Hanover, PA 17331
American Walking Pony Registry	Route 5, Box 88, Upper River Road	Macon, GA 31211
Welsh Pony Society of America	P.O. Box 2977	Winchester, VA 22601
Wild Horses of America Reg., Inc.	11790 Deodar Way	Reno, NV 89506
American Horse Shows Assoc.	598 Madison Avenue	New York, NY 10022
Dressage & Combined Training Assoc.	P.O. Box 12460	Cleveland, OH 44112
United States Dressage Federation Inc.	1212 O Street, P.O. Box 80668	Lincoln, NE 68501

Appendix B

Blister Beetle Poisoning

Blister beetles may be present in alfalfa hay cut primarily from areas of low humidity such as in the western and southwestern United States, and that cut after the middle of the summer. Ingestion of only a few blister beetles is fatal to the horse.

Blister beetles contain a toxin called cantharidin. This toxin is very stable and is present in beetles that are alive or in those that have been dead even for as long as several years. Cantharidin is an irritant which on any body surface produces severe inflammation and blisters, and therefore the name blister beetle. When the beetle is ingested, the toxin is absorbed and excreted in the urine, causing severe irritation and inflammation of the digestive and urinary tracts. The burning sensation that this causes upon urination after the ingestion of blister beetles is the basis behind their use in humans as an aphrodisiac called "Spanish Fly." The ingestion of several of these beetles by the horse results in necrosis and sluffing of the mucosal lining of the esophagus (see Glossary Fig. 2) and stomach, with severe reddening of the gastrointestinal and urinary tracts. Affected animals have colic, tenesmus, usually increased temperature, depression, increased heart and respiratory rates, dehydration, sweating, diarrhea, and frequently will continually play in water. If the horse survives for 24 hours, there is frequent urination of small amounts of urine. This, in the presence of dehydration and colic, are helpful diagnostic signs. The urine may or may not be blood-tinged but will nearly always contain occult blood that can be detected using Occultest Tablets. A decrease in the plasma calcium concentration is present for the first 48 hours after the onset of clinical signs but may not be present after this time. Death generally occurs within 48 to 72 hours after clinical signs are first present.

There is no specific treatment other than supportive care. The bowel should be evacuated to prevent the absorption of any toxin remaining in the stomach or intestine. Since cantharidin is fat soluble, mineral oil should be given by stomach tube. Analgesics should be given to decrease pain. Large quantities (20 to 40L) of fluid given intravenously are needed to correct dehydration and to induce diuresis and excretion of the toxin in a more dilute urine concentration. An extracellular replacement fluid such as Ringer's Lactate to which 10 ml of 23% calcium gluconate per liter has been added should be given.

Blister beetles vary from ¼ to 1 inch (0.6 to 2.5 cm) in length with a width about one-fourth their length (Appendix Fig. 1). Their head and neck are about one-half the width of their body and make up 20 to 25% of their length. Attached to the head are fairly long antennae. The beetles may be entirely black, black with orange stripes, gray or yellowish-tan with or without black spots. The larvae of blister beetles feed on grasshopper eggs and therefore their numbers may increase with increased grasshopper populations. Blister beetles have only one generation per year. In most areas, their adult population generally

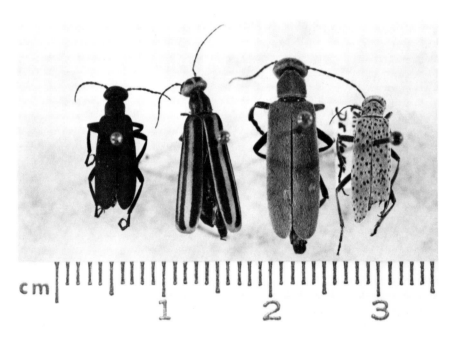

Appendix Fig. 1. Blister Beetles. They may be entirely black, black with orange stripes, gray or yellowish tan with or without black spots. They vary from ¼ to 1 inch (0.6 to 2.5 cm) in length with a width about one-fourth their length. Their distinctive neck, head and long antennae are helpful in differentiating them from other beetles. Ingestion of only a few of these beetles is fatal to the horse.

peaks in August. They tend to aggregate in groups and may not be evenly distributed in a field, and since they are highly mobile, can readily move to other areas. The potential hazard is accentuated when hay is cut, windrowed, and crimped all in one operation, since the beetles may be crushed in the crimping process and therefore cannot move out of the cut hay.

Blister beetles in most areas are not generally present in hay cut before the last part of July. The beetles are attracted by flowers and therefore are rarely present in nonflowering plants such as grasses. They are a problem only in alfalfa hay. Cutting alfalfa prior to flowering lessens the chances of it containing beetles. Cutting before flowering also increases the nutritional value of the hay. Waiting until after flowering greatly decreases its nutritional value. Spraying with the insecticide Sevin (a carbamate which is a cholinesterase inhibitor and is present in several commercial formulations including Sevin 80SP, Sevimol, and Sevin XLR) at an application rate of 1.5 lb of active ingredient/acre (1.5 kg/hectare) shortly before cutting may be necessary in infested fields. Hay sprayed with Sevin may be fed to animals with no detrimental effects and no waiting period after spraying. Sevin kills not only blister beetles but also grasshoppers, crickets, cutworms, armyworms, alfalfa caterpillars, and webworms.

Hay containing blister beetles, or from fields known to contain blister beetles, should not be fed to horses. Numerous small animals and birds are resistant to blister beetle poisoning. Ruminants, although susceptible to blister beetle poisoning, are more resistant to it than are horses.

Index

Page numbers in *italics* indicate illustrations; those followed by t indicate tables.

239